Neuropeptide Receptors

Colloquium Series on Neuropeptides

Editors
Lloyd D. Fricker, Ph.D.,
Professor, Department of Molecular Pharmacology, Department of Neuroscience, Albert Einstein College of Medicine, New York
Lakshmi A. Devi, Ph.D.,
Professor, Department of Pharmacology and Systems Therapeutics, Department of Neuroscience, Department of Psychiatry, Mount Sinai School of Medicine, New York.

Communication between cells is essential in all multicellular organisms, and even in many unicellular organisms. A variety of molecules are used for cell-cell signaling, including small molecules, proteins, and peptides. The term 'neuropeptide' refers specifically to peptides that function as neurotransmitters, and includes some peptides that also function in the endocrine system as peptide hormones. Neuropeptides represent the largest group of neurotransmitters, with hundreds of biologically active peptides and dozens of neuropeptide receptors known in mammalian systems, and many more peptides and receptors identified in invertebrate systems. In addition, a large number of peptides have been identified but not yet characterized in terms of function. The known functions of neuropeptides include a variety of physiological and behavioral processes such as feeding and body weight regulation, reproduction, anxiety, depression, pain, reward pathways, social behavior, and memory. This series will present the various neuropeptide systems and other aspects of neuropeptides (such as peptide biosynthesis), with individual volumes contributed by experts in the field.

Published titles
(To see published titles please go to the website, www.morganclaypool.com/page/lifesci)

Neuropeptide Receptors
Ivone Gomes, Jonathan H. Wardman, Steven D. Stockton Jr., and Lakshmi A. Devi
2013

Neuropeptides and Other Bioactive Peptides: From Discovery to Function
Lloyd D. Fricker
June 2012

Bioactive Peptides Produced by Limited Proteolysis
Antonio C. M. Camargo, Beatriz L. Fernandes, Lilian Cruz, Emer S. Ferro
2012

Peptide Biosynthesis: Prohormone Convertases 1/3 and 2
Akina Hoshino, Iris Lindberg
2012

Lectures under development

Carboxypeptidase E and D
Lloyd D. Fricker

Extracellular Processing and Degradation of Neuropeptides
Anthony Turner

Neurologically Active Peptides in Invertebrates
Lingjun Li

Neuropeptides in Circadian Physiology
Paul Taghert and Erik Herzog

Neuropeptide Receptors
Ivone Gomes, Jonathan H. Wardman, Steven D. Stockton Jr., and Lakshmi A. Devi

www.morganclaypool.com

ISBN: 9781615044689 print
ISBN: 9781615044696 ebook

DOI: 10.4199/C00082ED1V01Y201304NPE004

A Publication in the Morgan & Claypool Publishers series
COLLOQUIUM SERIES ON NEUROPEPTIDES
Lecture #4
Series Editor: Lloyd D Fricker, Albert Einstein College of Medicine, and Lakshmi A. Devi, Mount Sinai School of Medicine

Series ISSN 2166-6628 Print 2166-6636 Electronic

Neuropeptide Receptors

Ivone Gomes, Jonathan H. Wardman, Steven D. Stockton Jr., and Lakshmi A. Devi
Departments of Pharmacology and Systems Therapeutics, Psychiatry and Neuroscience and The Friedman Brain Institute, Mount Sinai School of Medicine

COLLOQUIUM SERIES ON NEUROPEPTIDES #4

MORGAN & CLAYPOOL LIFE SCIENCES

ABSTRACT

Neuropeptides mediate their effects by binding and activating receptors that are responsible for converting these extracellular stimuli into intracellular responses. Most neuropeptides interact with G protein-coupled receptors that transduce the signal by activating heterotrimeric G proteins leading to alterations in second messenger systems to amplify the signal and elicit the intracellular response. In this review, we describe the general structure of G protein-coupled receptors including the information obtained from crystal structure determination that has given an insight into the activation mechanism of these receptors. In addition, we summarize the components of the signal transduction system (including G proteins, effectors and second messengers generally activated by the neuropeptide receptors). Using select examples of neuropeptide-receptor systems, we highlight the neuropeptides and corresponding receptors involved in modulation of pain and analgesia, body weight regulation, and hormonal regulation. Finally, we discuss the enzyme-linked tyrosine kinase receptors activated by growth factors and discuss the emerging concepts in targeting neuropeptide receptors for the identification of novel therapeutics targeting these systems.

KEYWORDS

G protein-coupled receptor, heptahelical receptor, 7 transmembrane receptor, enkephalin, morphine, opioid peptide receptor, neuropeptide Y receptor, GPCR heteromers

Contents

Abbreviations

2-AG	2-arachidonylglycerol
5HT2A	serotonin 2A
AC	adenylyl cyclase
ACE	angiotensin-converting enzyme
ACTH	adrenocorticotrophic hormone
ADP	adenosine diphosphate
AgRP	agouti related peptide
AGS	activators of G protein signaling
AKAPs	A-kinase anchoring proteins
Ala	alanine
AMP	adenosine monophosphate
AngII	angiotensin II
Arg	arginine
Asn	asparagine
Asp	aspartic acid
AT1R	AT1 angiotensin receptor
β-FNA	β-funaltrexamine
BDNF	brain-derived neurotrophic factor
CAMK	Ca^{+2}/calmodulin dependent kinase
cAMP	cyclic AMP
CART	cocaine- and amphetamine-regulated transcript
CCK	cholecystokinin
cGMP	cyclic guanosine monophosphate
CGRP	calcitonin gene related peptide
CNBD	cyclic nucleotide binding domain
CNG	cyclic nucleotide gated ion channel
CPD	carboxypeptidase D
CPE	carboxypeptidase E
CPON	C-terminal peptide of neuropeptide Y
CPP	conditioned place preference
CREB	cAMP response-element binding protein

CRHR1	corticotropin releasing hormone receptor 1
CTX	cholera toxin
Cys	cysteine
DAG	diacylglycerol
DAMGO	[D-Ala^{2}, N-MePhe^{4}, Gly-ol]-enkephalin
ECE-1	endothelin-converting enzyme 1
ER	endoplasmic reticulum
ERK1/2	extracellular signal-regulated kinases 1/2
GABA	γ-aminobutyric acid
GAP	GTPase activating protein
GAP43	growth associated protein 43
GDP	guanosine diphosphate
GEF	guanine nucleotide exchange factor
GIRK/Kir3	G protein-gated inwardly rectifying potassium channel
Gln	glutamine
GLP-1	glucagon-like peptide 1
GLUT4	glucose transported type 4
Gly	glycine
GPCR	G protein-coupled receptor
GPI	glycerophosphatidylinositol
GRK	G protein receptor kinase
GTP	guanosine triphosphate
FSH	follicle stimulating hormone
His	histidine
IGF-1	insulin-like growth factor-1
Ile	isoleucine
IP3	inositol-1,4,5-triphosphate
JAK	Janus kinase
JNK/SAPK	c-Jun N-terminal kinases/stress-activated protein kinase
kDA	kilo Daltons
LDCV	large dense-core vesicles
LepR	leptin receptor
Leu	leucine
LIMK	LIM domain kinase
Lys	lysine
MAPK	mitogen activated protein kinase
MC	melanocortin

MEK1/2	mitogen activated protein kinase kinase1/2
Met	methionine
MSH	melanocyte stimulating hormone
NAD	nicotinamide adenine dinucleotide
NGF	nerve growth factor
NK	neurokinin
NOP	nociceptin/orphanin FQ receptor
NPY	neuropeptide Y
NT-3	neurotrophin 3
NT-4	neurotrophin 4
NTS	nucleus tractus solitarius
PAK1	p21-activated kinase
PAM	peptidylglycine α-amidating monooxygenase
PAR-2	protease activated receptor-2
PC	prohormone convertase
PDE	phosphodiesterase
PDK-1	phosphoinositide-dependent kinase-1
PFC	prefrontal cortex
Phe	phenylalanine
PI-3-K	phosphoinositide-3-kinase
PIP_2	phosphatidylinositol 4,5-bisphosphate
PKA	protein kinase A
PKC	protein kinase C
PLCβ	phospholipase Cβ
POMC	proopiomelanocortin
PPTA	preprotachykinin A
PPTB	preprotachykinin B
Pro	proline
PTH	parathyroid hormone
PTX	pertussis toxin
PVN	paraventricular nucleus
RGS	regulators of G-protein signaling
RhoGEF	Rho guanine nucleotide exchange factor
RNA	ribonucleic acid
STAT	signal transducer and activator of transcription
TFMPP	3-trifluoromethylphenylpiperazine
TGF-β	transforming growth factor β

TGN	*trans*-Golgi network
THC	Δ^9-tetrahydrocannabinol
TM	transmembrane
Trp	tryptophan
TSH	thryroid stimulating hormone
Tyr	tyrosine
VDCC	voltage-dependent calcium channel

CHAPTER 1

Overview of Neuropeptide Receptors

Receptors are signal transducing proteins that respond to a specific stimulus in the cell's environment by eliciting an intracellular response. They convert the information from an extracellular signaling molecule into either electrical activity or chemical changes in the cell. This change is then thought to cause secondary events that produce a cellular response. In addition to signal transduction, receptors do two other things: signal amplification and signal processing. Amplification occurs when the binding of a ligand to a receptor induces a change in the intracellular levels of second messengers. Signal processing occurs when receptor activation leads to the activation of a variety of intracellular events.

The simple concept that a neuropeotide receptor is a protein that responds to one specific neuropeptide arose only in the last century. In the early 20th century, it was generally assumed that the communication between neurons was electrical. Tissue staining techniques were inadequate and hence the knowledge of neurons and neurotransmission was limited. Camillo Golgi, while experimenting with metal impregnation of tissue, discovered a method of staining nervous tissue [1]. In late 1880's Ramon y Cajal used Golgi's silver nitrate preparations to provide a detailed description of neuronal structures including axons, dendrites and dendritic spines. This led him to develop the electrochemical theory of neurotransmission (with chemical signaling at the synapse) that contrasted with the electrically coupled neurotransmission proposed by Golgi [1]. In 1921 Otto Loewi carried out key experiments with frog hearts that documented chemical neurotransmission. He showed that a substance released from the electrically stimulated donor heart could affect the beating of a recipient heart and suggested that the donor heart released a soluble chemical (termed Vagustoff) later identified as acetylcholine [2]. This experiment demonstrated that electrical signals are converted to chemical signals at synapses. This also demonstrated the presence of a protein 'receptor' responsible for receiving the chemical signal and eliciting a response. In the early 1900s, studies with exogenous compounds led John Langely to conclude the presence of a 'receptive substance' at the nerve endings and this was subsequently termed 'receptor' [3]. At about the same time Paul Ehrlich proposed the existence of 'chemoreceptors' for drugs in recipient tissues [4]. While these experiments essentially defined the concept of a neurotransmitter and its receptor, expansion and rethinking of the definition of transmitters came with the discovery of exogenous compounds (e.g., nicotine, morphine) that have the ability to elicit post-synaptic effects and alter neurotransmission.

The availability of a bioassay that could easily detect the activation of receptors that were targets of morphine led to the identification of enkephalins by Kosterlitz and colleagues [5]. Enkephalin receptors were the first neuropeptide receptors to be well characterized [6, 7]. Subsequently, a large number of neuropeptides with diverse activities were identified and shown to play a role in physiological functions ranging from pain and analgesia to feeding and bodyweight regulation. The majority of these peptides activate *G protein-coupled receptors (GPCRs)*, which mediate many of their effects by activation of heterotrimeric G proteins, producing cellular effects by inducing the production of second messengers [8–10]. In a subset of cases, the peptides are able to activate *enzyme-linked receptors* in the nervous system, which respond to external stimulation by forming dimers that result in the activation of intracellular enzymes.

In the following chapters we describe the general structure of GPCRs (Chapter 2) as well as the components of the signal trasduction system (G proteins, effectors and second messengers) (Chapter 3). We also describe how neuropeptides are processed from larger precursor proteins (Chapter 4) and provide examples of classic neuropeptide-receptor systems (Chapter 5). In addition, we describe the current advances in the field that could lead to identification of novel therapeutic targets (Chapter 6).

CHAPTER 2

G Protein-Coupled Receptors: General Structure & Function

2.1 INTRODUCTION

Neuropeptides exert their effects through the activation of GPCRs. The latter constitute the largest family of cell surface membrane receptors (~950 unique receptors) in the human genome that convert extracellular signals into a physiological response [11, 12]. Given that the signaling pathways mediated through GPCR activation regulate a wide range of critical biological functions, this makes GPCRs a significant target for as many as 30% of all pharmacological agents currently in use [13]. GPCRs are integral membrane proteins characterized by a 7-transmembrane (TM) topology. This 7-TM topology is evolutionarily conserved from yeast to man and widely used in nature to transduce signals from a diverse array of endogenous ligands [14]. These include small molecules such as calcium, amino acid-derived neurotransmitters (dopamine) and biogenic amines (norepinephrine, 5-hydroxytryptamine, histamine), to larger entities such as neuropeptides (e.g., enkephalin, neuropeptide Y) and proteins (e.g., follicle stimulating hormone). Despite the diversity in the nature of these ligands, the mechanism of activation appears to be fairly similar. Binding of the ligand to extracellular regions of the receptor leads to conformational changes of the TM regions; this movement of the TMs in relation to one another results in activation of intracellular signaling. Early studies examining the signaling by GPCRs used the classic "lock and key" model to describe receptor activation. According to this model the receptor exists in two states—an 'off' state and an 'on' state; binding of the agonist (acting as the 'key') would lead to unlocking of the receptor from an 'off' to an 'on' state. Such a model predicted the receptor to be a relatively rigid molecule that could exist in only two conformations. In contrast to this idea, it is becoming increasingly clear that the receptor is an agile molecule that can exist in multiple conformations; ligands based on their physical-chemical properties can stabilize different receptor conformations. Based on this, ligands have been classified as agonists that stabilize the active conformation, antagonists that compete for agonist binding and inverse agonists that stabilize the inactive conformation. Recent crystal structure studies are beginning to reveal the mode of activation of GPCRs at the molecular level. In the following sections we describe the topology and classification of GPCRs and structural information obtained from the recent crystallization of peptidergic GPCRs.

2.2 TOPOLOGY AND CLASSIFICATION OF G PROTEIN-COUPLED RECEPTORS

GPCRs can be organized into an extracellular domain, a transmembrane (TM) core and an intracellular domain [11]. The extracellular domain comprises the N-terminal tail and the extracellular loops (e1, e2, e3), the hydrophobic TM core is formed by the bundling of the seven α-helical regions of the receptor (TM1 to TM7), while the intracellular domain includes the C-terminal tail and the intracellular loops (i1, i2, i3). GPCRs exhibit the greatest sequence homology within the TM domains. Conversely, greatest diversity is observed in the N- and C- termini as well as in the intracellular loop connecting TM5 and TM6 [15]. GPCRs, as the name implies, are associated with heterotrimeric G proteins (Gαβγ); the latter initiate a series of signal transduction events following their activation after the binding of a receptor agonist [16, 17]. The different heterotrimeric G proteins and the signaling cascades they activate are described in Chapter 3. Although GPCRs share these common features, they differ widely in their amino acid sequences. Thus, the GPCR superfamily is currently classified into six broad classes (A to F) based upon the presence of unique structural elements shared by broad groups of receptors. Among these, Class A, B and C are the most prevalent [18]; Class D and E represent families found predominantly in fungi and Class F GPCRs include the frizzled/smoothened receptors [11, 19]. The characteristics of family A, B, and C receptors are described below.

2.2.1 CLASS A GPCRS

Class A GPCRs, also known as Class 1 or rhodopsin-like receptors, represent the largest class of GPCRs, accounting for almost 85% of all known GPCRs [20]. While more than half of Class A GPCRs are predicted to be olfactory receptors, the other members of this class bind a diverse array of ligands that includes light (photons), biogenic amines, peptides and hormones [16, 21]. Several orphan GPCRs for which no ligand, either endogenous or exogenous, has yet been identified, also belong to this class of receptors [16, 21]. The topology of Class A GPCRs is remarkably conserved across all members of the family in spite of the limited sequence homology among them (Fig. 2.1). Studies show that the N-terminal domain of Class A GPCRs plays a role in ligand binding and receptor activation [16, 21]. The TM core of the receptor conveys the conformational changes induced by ligand binding to the extracellular side of the receptor to the intracellular domain of the receptor. The intracellular domain, in turn, converts ligand-induced conformational changes, either directly (through interactions with heterotrimeric G proteins) or indirectly (through interactions with scaffolding proteins), into intracellular signals via a variety of effector molecules (described in Chapter 3). In this context, the third intracellular loop of several Class A GPCRs has been shown to contain sites for the binding of a number of effectors [22]. In addition, post-translational modifications of the C-terminal domain (including palmitoylation and phosphorylation) modulate not only the coupling of G proteins to the receptor but also its activation state [23, 24].

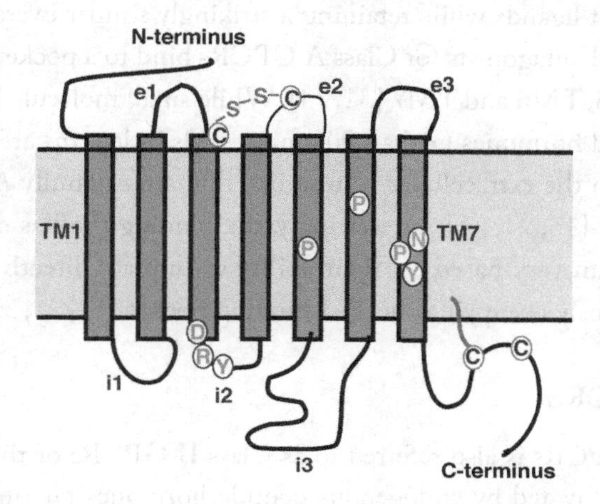

Figure 2.1. **Characteristics of Class A GPCRs.** Class A GPCRs are characterized by the presence of (i) a disulfide bond between conserved Cys residues in TM3 and e2; (ii) conserved Pro residues in TM5, TM6 and TM7; (iii) an E/DRY motif at the bottom of TM3; (iv) an NPXXY motif in TM7; and (v) a Cys residue in the C-terminal region that can be palmitoylated.

A feature common to most Class A GPCRs is the presence of a conserved cysteine (Cys) residue in TM3 and another in e2 that are thought to be involved in the formation of a disulfide bond and hence could play a role not only in the structural integrity but also in the functional activity of the receptor [25]. In addition, the intracellular C-terminal domain of Class A receptors has Cys residues proximal to the TM core that are thought to undergo dynamic palmitoylation-depalmitoylation events [26]. The presence of highly conserved proline residues within TM5, 6 and 7 introduce kinks into the helical structure of the TMs that are thought to be required to maintain the structural and/or functional integrity of the receptor [27]. Finally, Class A GPCRs are characterized by the presence of an E/DRY motif at the base of TM3 and a NPXXY motif in TM7 [28, 29]. Studies show that for some Class A GPCRs the E/DRY motif plays a role in keeping the receptor in the ground state while for other receptors it is involved in recognition/coupling to G proteins [28]. The NPXXY motif has been implicated in GPCR activation and in modulating the conformational switch occurring between multiple receptor activation states [29].

The ligand binding characteristics of Class A receptors have been among the most heavily studied aspects of GPCR research to date. A variety of experimental methodologies including computational models contributed to the current knowledge about ligand binding and receptor activity [30]. The ligand binding pocket of rhodopsin has been used as a template for understanding the ligand binding characteristics of Class A receptors [27, 30, 31]. It is thought that subtle differences within the TM helices of different members of this family underlie the ability of these receptors to bind such a diversity of ligands while retaining a strikingly similar overall surface topology [27, 30–32]. Both agonists and antagonists for Class A GPCRs bind to a pocket outlined by amino acid residues from TM3, TM5, TM6 and TM7 [31, 33]. While small molecule ligands bind to this TM pocket, large peptides and hormones (> than 90 amino acids in length) are recognized by high-affinity binding sites within the extracellular N-terminal domain of family A GPCRs including the extracellular loop regions [16, 21, 31]. Interestingly, the binding profiles of smaller peptides (<40 amino acids in length) can vary based on their ability to interact directly with extracellular loop regions while simultaneously occupying the TM binding pocket [16, 21, 31].

2.2.2 CLASS B GPCRS

The Class B family of GPCRs is also referred to as Class II GPCRs or the secretin receptor family. These receptors are activated by endogenous peptide hormones ranging in size from ~30-140 amino acid residues. Examples include secretin, vasoactive intestinal peptide, pituitary adenylate cyclase-activating polypeptide and glucagon. Class B GPCRs share a number of common structural features (Fig. 2.2). The extracellular N-terminal domain is moderately sized (approximately 100-160 residue long), and has a number of conserved Cys residues [11, 34]. In addition, the N-terminus of Class B receptors has a signal peptide that facilitates efficient cell surface expression of the receptor [34]. Like Class A GPCRs, Class B GPCRs have a number of conserved proline (Pro) residues within the TM domains although they are not at the same positions as those in Class A GPCRs [35]. Another prominent feature of the topography of Class B GPCRs is their basal level of glycosylation, with consensus sequences for Asn-linked glycosylation being found in the N-terminus, and sometimes within extracellular loops [30, 34]. Site-directed mutagenesis studies show that the N-terminal domain of Class B receptors close to TM1 is essential for ligand binding although the TM domains and extracellular loops provide information that determines the specificity of receptor-ligand interactions [34]. The N-terminal ligand binding domain has 3-4 conserved Cys residues and two conserved tryptophan (Trp) residues [34].

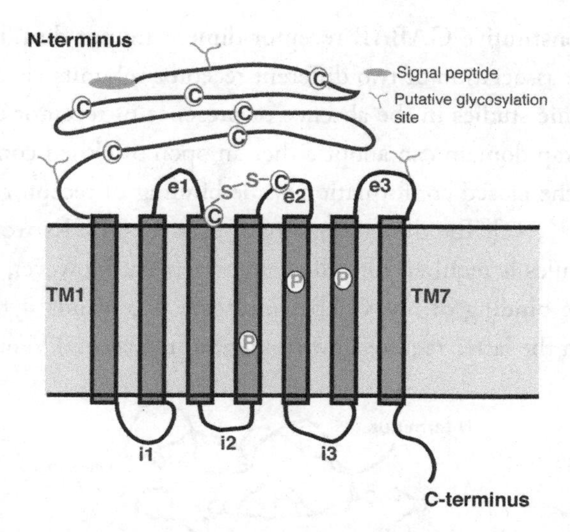

Figure 2.2. **Characteristics of Class B GPCRs.** Class B GPCRs are characterized by the presence of (i) a signal peptide in the N-terminal region; (ii) conserved Cys residues in the N-terminal region; (iii) a disulfide bond between conserved Cys residues in TM3 and e2; (iv) conserved Pro residues in TM4, TM5 and TM6; and (v) consensus sequence for Asn glycosylation in the N-terminal region and extracellular loops.

2.2.3 CLASS C GPCRS

Class C family GPCRs also known as metabotropic glutamate receptor (mGluR) family, includes in addition to metabotropic glutamate receptors, receptors for GABA (GABAB), calcium-sensing receptors as well as taste receptors [36]. Class C GPCRs are characterized by the presence of a Venus flytrap domain, a Cys-rich domain, a heptahelical TM domain and an intracellular very long C-terminal tail [36, 37] (Fig. 2.3). The N-terminal domain of Class C GPCRs is organized into two distinct extracellular "lobes" forming a Venus flytrap domain, with the space between the lobes forming the ligand-binding site for the receptor. The Cys-rich domain links the Venus flytrap domain to the heptahelical TM domain via a conserved disulfide bridge and appears to be necessary for the functioning of these receptors [36, 37]. Conformational changes induced by ligand binding to the Venus flytrap domain are transmitted to the TM spanning domains and this promotes coupling to intracellular G proteins and activation of signaling transduction pathways [36, 37]. In addition, class C GPCRs, particularly glutamate and GABAB receptors, contain a very highly conserved NEAK/NDSK motif that is found along the length of the i3 loop [36, 37].

Early studies demonstrated that Class C GPCRs function as obligate dimers. Both in the case of metabotrobic glutamate receptors and of calcium-sensing receptors a disulfide bridge involving a Cys residue in the Venus flytrap domain was found to contribute toward dimer formation

[38, 39]. Interestingly, constitutive GABAB receptor dimers are not disulfide linked; in this case dimerization involves the association of two different receptor subunits via coiled-coil interactions [40]. X-ray crystallographic studies in the absence or presence of receptor agonists or antagonists show that the Venus flytrap domain can adopt either an open or closed conformation. Binding of receptor agonists favors the closed conformation while binding of receptor antagonists favors the open conformation [36, 41, 42]. The dimeric nature of class C GPCRs would suggest that binding of two agonist molecules is required for full receptor activity. However, this is not the case for GABAB receptors where binding of one GABA molecule to a Venus flytrap domain elicits full receptor activity although the latter requires involvement of the second Venus flytrap domain [43].

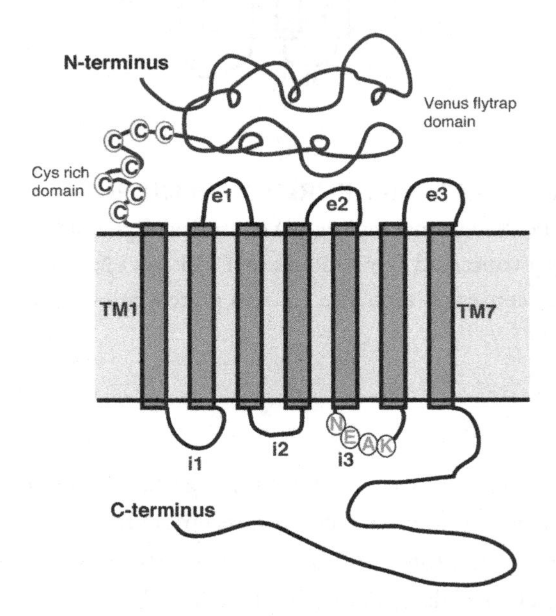

Figure 2.3. **Characteristics of Class C GPCRs.** Class C GPCRs are characterized by the presence of (i) a N-terminus organized into a Venus flytrap-like domain; (ii) a Cys-rich N-terminal domain; (iii) a NEAK/NDSK motif in i3 loop.

2.3 CONFORMATIONAL AND STRUCTURAL CHANGES ASSOCIATED WITH RECEPTOR ACTIVATION

Given that GPCRs share a common basic architecture comprising of a 7 TM α-helical bundle and that as a group they recognize a wide variety of ligands to induce a specific physiological response, it is important to understand not only the mechanisms driving signal transduction by these receptors but also what drives receptor ligand selectivity and specificity. This could be achieved through the use of biophysical and biochemical approaches in combination with high-resolution GPCR structures. However, despite intensive efforts, until very recently, the only available high-resolution structure for a GPCR was that of rhodopsin. The limitations to obtaining high-resolution GPCR structures included low endogenous receptor expression in tissues, difficulties in expressing the receptor at high levels in heterologous cells, structural instability following detergent solubilization as well as the fact that the receptors exist in multiple conformational states. Recent approaches aimed at stabilizing a GPCR into a single conformational state have permitted the determination of high-resolution structures for a number of GPCRs [44–47]. In the case of opioid and chemokine receptors the approach involved the replacement of the disordered third intracellular loop with T4 lysozyme [48]. In the following sections we describe the information derived from the crystal structure of peptidic GPCRs particularly with regard to the plasticity of the TM hydrophobic core and how the latter is modulated by ligand binding.

2.3.1 THE CRYSTAL STRUCTURE OF CXCR4 RECEPTOR

CXCR4, the receptor for the chemokine CXCL12, was the first peptide GPCR to be crystallized. The crystal structures for the human CXCR4 chemokine receptor were generated using the T4 lysozyme fusion protein strategy originally described for the crystallization of β2-adrenergic receptors [46]. For this, T4 lysozyme was inserted between TM5 and TM6 replacing the third intracellular loop. In addition, the constructs contained a thermostabilizing L125W mutation [46].

The 2.5 Å resolution crystal of CXCR4 in complex with the antagonist IT1t (a small drug like isothiourea derivative) reveals several differences from the structures of two other family A GPCRs (β_2-adrenergic and A_{2A} adenosine receptors) specially with regard to the organization of the TM domains. For example, the extracellular end of TM1 is shifted toward the central axis of the receptor as compared to β_2-adrenergic and A2A adenosine receptors [46]. TM2 makes a tighter helical turn at Pro92 resulting in a ~120° rotation of its extracellular end compared to other GPCR structures [46]. Both the intracellular and extracellular tips of TM4 in CXCR4 deviate substantially from their consensus positions in other GPCRs [46]. The extracellular end of TM5 in CXCR4 is about one turn longer and that of TM7 is two helical turns longer than in other GPCR structures [46]. The disulfide bridges between Cys28 and Cys274 and between Cys109 in TM3 and Cys186 in the e2 loop are critical for ligand binding [49] and function by constraining the e2 loop and the N-terminal region between residues 26–34, thereby shaping the entrance to the ligand binding pocket [46]. Surprisingly, TM7 is about one turn shorter at the intracellular side, ending right after the conserved NPxxY motif, and lacks the short H8 helix as well as the putative palmitoylation site [46].

A comparison of the crystal structure of CXCR4 complexed to IT1t with that complexed with the cyclic peptide inhibitor (CVX15) shows a significant overlap between both structures.

IT1t, occupies part of the binding pocket defined by side chains from TM1, TM2, TM3 and TM7 but makes no contact with TM4, TM5 and TM6 [46]. The nitrogens of the symmetrical iso-thiourea group in the IT1t ligand are protonated and one of them forms a salt bridge with the side chain of Asp97 in the receptor [46]. Connected by a short flexible linker, the imidazo-thiazole ring system in the IT1t ligand contacts TM7 by making a salt bridge with Glu288 [46]. In the case of the bulky 16-residue cyclic peptide inhibitor, CVX15, it fills most of the binding pocket defined by the side-chains from TM1, TM2, TM3 and TM7. The N-terminal region of the peptide backbone from Arg1 to Cys4 forms hydrogen bonds with the receptor, adding a partial third strand to the e2 loop β-hairpin [46]. The core specific interactions are formed by the two Arg residues at the peptide N-terminus: Arg1 makes polar interactions with Asp187 on the receptor while Arg2 interacts with Thr117 and Asp171 and may form an additional hydrogen bond with His113 depending on its protonation state [46]. The bulky naphthalene ring on the peptide is anchored in a hydrophobic region bordered by TM5 [46]. Arg14 on the peptide makes a salt bridge with Asp262, and an intramolecular hydrogen bond with the peptide Tyr5 side chain, which in turn makes hydrophobic contacts with TM5 side chains [46]. Finally, the C-terminal d-Pro on the peptide is buried in the pocket next to the N-terminus of the peptide, making a water-mediated interaction with Asp288 side chain of the receptor [46].

An interesting aspect of all CXCR4 crystal structures is that they exhibit a similar parallel, symmetric dimer indicating that the receptors exist as dimers [46]. In the context of CXCR4 dimers bound to the small ligand, IT1t, the protomers interact only at the extracellular side of TM5 and TM6, leaving at least a 4 Å gap between the intracellular regions, which is presumably filled by lipids [46]. However, CXCR4 dimers bound to the bulky peptide, CVX15, are stabilized by interactions at the intracellular ends of TM3, TM4, and i2 loop involving mostly hydrophobic interactions [46]. It is likely that binding of the bulky CVX15 peptide induces a small tilt in the extracellular part of TM5, which brings the intracellular parts of opposing protomers into close contact. This could explain the cooperative binding observed with certain CXCR4 ligands, as well as the effects of allosteric modulators [46].

2.3.2 THE CRYSTAL STRUCTURE OF THE μ OPIOID RECEPTOR

A 2.8Å crystal structure of the μ opioid receptor in complex with the irreversible antagonist, β-funaltrexamine (β-FNA) was obtained using the T4 lysozyme fusion protein strategy [45]. The intracellular side of the μ opioid receptor resembles that of rhodopsin with respect to the relative positions of TM3, TM5 and TM6 [45]. There is a disulfide bridge between Cys140 and Cys217 connecting the e2 loop to TM3. Similar to β-adrenergic receptors, the DRY motif in the μ opioid receptor does not form an ionic lock but interacts with the i2 loop via polar hydrogen bonds [45]. β-FNA makes contacts with TM3, TM5, TM6, TM7 and the side chain of K233 is the site of covalent attachment [45]. The closest published structure similar to the μ opioid receptor is that of the CXCR4 chemokine receptor (described in 2.3.1). Comparison of the two structures shows that the μ opioid receptor ligand, β-FNA, binds much more deeply within the receptor binding pocket compared to the CXCR4 antagonist [45]. In the binding pocket of μ opioid receptors there are 14 residues within 4Å of β-FNA, nine of which make direct interactions with this ligand and are found to be conserved in δ and κ opioid receptors [45]. In TM3, D147 makes charge-charge interactions with the amine moiety of β-FNA and also hydrogen bonds with Y326 in TM7. H297 in TM6 interacts with the aromatic ring of β-FNA via water molecules that are well positioned to form a hydrogen-bonded network between H297 and the phenolic hydroxyl of β-FNA [45].

The lattice structure of μ opioid receptors shows alternating aqueous and lipidic layers with the receptors arranged in parallel dimers tightly associated through TM5 and TM6 [45]. In addition, limited interdimeric contacts through TM1, TM2 and helix 8 were observed between adjacent dimers [45]. This supports biochemical evidence accumulated over the last decade showing that μ opioid receptors form homodimers [50–52]. In addition, μ opioid receptors can form heterodimers with a number of family A GPCRs (see Chapter 5).

2.3.3 THE CRYSTAL STRUCTURE OF THE κ OPIOID RECEPTOR

The 2.9Å resolution crystal structure of the human κ opioid receptor in complex with the selective antagonist JDTic has been obtained by taking advantage of the T4 lysozyme strategy [47]. Unlike the crystal structures of non-rhodopsin receptors that indicate the presence of more than one disulfide bond, the human κ opioid receptor has only one between Cys131 and Cys210 that bridges the e2 loop to the end of TM3. This disulfide bond is also present in the solved structures of other family A GPCRs [47]. Interestingly, the e3 loop of the κ opioid receptor is disordered [47]. With regard to the DRY motif present in most class A GPCRs that is thought to form the "ionic lock" that stabilizes the inactive conformation of the receptor, the human κ opioid receptor lacks the acidic residues present in this motif but the Arg residue forms a hydrogen bond with Thr273 on TM6 to stabilize the inactive conformation of the receptor [47].

As seen with the structure of CXCR4, the human κ opioid receptor binding pocket is comparatively large and partially capped by the e2 β-hairpin loop, although it is much narrower and deeper than observed in CXCR4 [47]. The differences in the shape of the binding pocket could be due to a ~ 4.5 Å inward shift of the extracellular tip of TM6 in κ opioid receptor as compared to CXCR4 as well as to differences in the side chains lining the pocket [47]. The JDTic ligand reaches deep into the pocket to form ionic interactions with the side chain of Asp 138. The protonated amines in both piperidine and isoquinoline moieties of JDTic form salt bridges with the side chain of Asp138 [47]. This Asp residue is conserved in all aminergic GPCRs and in all opioid receptor subtypes, and may play a role in anchoring positively charged κ receptor ligands [47]. The isopropyl group of JDTic reaches deep into the binding pocket to form a hydrophobic interaction with the side chain of Trp287; this Trp residue is thought to be a key part of the activation mechanism in many class A GPCRs [47]. Thus, JDTic fits tightly to the bottom of the binding cleft by forming an array of ionic, polar and hydrophobic interactions with the receptor [47]. Interactions between JDTic and four residues in the κ opioid receptor (Val108, Val118, Ile294 and Tyr312) contribute to the receptor subtype selectivity of this ligand [47].

The asymmetric crystal unit of the κ opioid receptor consists of two receptors forming a parallel dimer. The dimer interface is formed through contacts among TM1, TM2 and helix 8 [47]. The extracellular domain of the human κ opioid receptor is very similar to that of the CXCR4 receptor while the TM domain is similar to that of β2-adrenergic and dopamine D3 receptors [47]. The i2 loop of individual κ opioid receptors in the asymmetric parallel dimer adopts slightly different structures suggesting a conformational plasticity in this region of the receptor [47].

2.3.4 THE CRYSTAL STRUCTURE OF THE δ OPIOID RECEPTOR

The 3.4Å resolution crystal structure of the mouse δ opioid receptor in complex with the selective antagonist naltrindole has been obtained by taking advantage of the T4 lysozyme strategy [44]. The δ opioid receptor exhibits remarkable conservation of backbone structure with other opioid receptors even in regions with low sequence homology [44]. The antagonist naltrindole sits in an exposed binding pocket that is similar in shape to that observed for the μ and κ opioid receptors [44]. The e2 loop of the δ opioid receptor, like that of μ and κ opioid receptors, also has a β-hairpin structure [44]. The extracellular half of TM1 of the δ opioid receptor is more similar to μ and CXCR4 receptors than to κ opioid receptors [44].

The position of the antagonist naltrindole in the binding pocket of the δ opioid receptor is slightly shifted relative to the position of the irreversible antagonist β-FNA bound to the μ opioid receptor although all major interactions are present in both structures [44]. Leu300 in the receptor is in contact with the indole group of naltrindole and is responsible for naltrindole's selectivity for δ opioid receptors.

The crystal structure of opioid receptors suggests that the concept of "message–address" for an opioid ligand could be a direct consequence of the structure of the receptor [53]. The lower portion of the binding pocket of opioid receptors, that is well conserved in terms of sequence and structure, would recognize the "message" portion of the ligand while the upper portion of the binding pocket, which is divergent among receptor subtypes and rich in selectivity determinants, would comprise the "address" portion of the receptor ligand [53]. In the context of the δ opioid receptor, the core morphian group of naltrindole (representing the "message" portion of the antagonist) fits into the lower binding pocket while the indole portion of naltrindole (comprising the "address" portion of the ligand) fits into the upper binding pocket.

In contrast to the μ and κ opioid receptor crystal lattice of two parallel dimeric interfaces, the δ opioid receptor crystallizes as an anti-parallel dimer [44]. This could be due to either differences in crystallization conditions, T4 lysozyme arrangement, sequence differences in the protein, and/or subtle differences in the structures stabilized by the different ligands.

2.3.5 THE CRYSTAL STRUCTURE OF THE NOCICEPTIN/ORPHANIN FQ RECEPTOR

The 3Å resolution crystal structure of the human nociceptin/orphanin FQ receptor (NOP) in complex with a small peptidomimetic antagonist, Compound-24 (C-24), has been obtained by replacing N-terminal residues with thermostabilized apocytochrome b_{562}RIL and by deletion of 31 C-terminal residues [54]. Comparison of the NOP structure with that of opioid receptors shows intriguing variations in the 7TM domain. These include (i) the presence of five Pro-induced kinks that play an important role in determining the shape of the ligand binding pocket [54]; (ii) a shift in the extracellular tip of TM5 by more that 4 Å, compared to the crystal structures of μ and κ opioid receptors which creates a gap between TM4 and TM5 leading to an expansion of the ligand binding pocket [54]; and (iii) a tilt in the extracellular tip of TM6 and TM7 toward the ligand binding pocket compared to CXCR4 [54].

In the case of the extracellular loops, the e1 and e2 loops of NOP are structurally similar to that of CXCR4 and κ opioid receptors [54]. Similar to opioid receptors, the e2 loop of NOP forms a β-hairpin structure that is anchored to the extracellular tip of TM3 by a conserved disulfide bond [54]. Moreover, like in κ opioid receptors, the e2 loop of NOP is rich in glutamate and aspartate residues making not only this loop but also the entrance to the ligand binding pocket acidic; however, the e2 loop is two residues shorter than the one present in κ opioid receptors [54].

The NOP crystal structure shows that the antagonist C-24 binds deep within the ligand binding pocket with the protonated nitrogen of the C-24 piperidine forming a salt bridge with Asp130 (this residue is conserved in opioid receptors) [54]. In addition, the linked benzofuran/piperidine rings in C-24 are buried in a hydrophobic pocket created by residues from TM3, TM5 and TM6 [54]. The benzofuran group in C-24 is sandwiched between Met134 and Tyr131; interestingly the side-chain of Met134 adopts a more buried conformation compared to that observed in κ opioid receptors which permits the deeper penetration of the C-24 ring system into the binding pocket [54]. Finally, the carbonyl group in C-24 that is adjacent to the pyrrolidone ring forms a hydrogen bond with Gln107, a residue that also forms a hydrogen bond with Tyr309; both Gln107 and Tyr309 are present in κ opioid receptors although in different conformations [54]. Comparisons of the NOP structure with that of opioid receptors show that three residue positions in the binding pocket of NOP (Ala 216, Gln280 and Thr305) that differ from opioid receptors (Lys216, His280, Ile305) play a major role in shaping the ligand binding pocket.

2.4 DOMAINS INVOLVED IN GPCR DIMERIZATION/ HETEROMERIZATION

For the last decade several lines of evidence have demonstrated that GPCRs can associate to form higher order complexes. These receptor-receptor associations, particularly heterodimerization, modulate not only the binding but also signaling and trafficking properties of individual receptors. Studies examining the domains involved in GPCR homodimerization/heterodimerization have implicated a role for extracellular, TM, and/or C-terminal regions. Moreover, these associations could involve either covalent (disulfide bonds) and/or non-covalent (hydrophobic) interactions.

Early studies examining dimerization of family C receptors have demonstrated the involvement of disulfide bonds between conserved cysteine residues in the extracellular domain of Ca^{+2}-sensing receptors [55, 56]. Moreover, X-ray crystallographic analysis of the extracellular ligand-binding domain of the metabotropic glutamate receptor revealed that it forms a disulfide-linked homodimer where each monomer has a bi-lobed structure [36]. Biochemical studies have implicated the extracellular domain in the dimerization of bradykinin B1 (a family A GPCR) and Ig-Hepta (a family B GPCR) receptors [57, 58].

The involvement of the C-terminal domain in GPCR dimerization has been suggested by studies carried out with $GABA_B$ and δ opioid receptors. In the case of $GABA_B$ receptors, heterodimerization between $GABA_BR1$ and $GABA_BR2$ was found to be necessary for the formation of a functional receptor [59, 60]. Studies show that the C-terminal tails of these receptors have a coiled-coil motif and that interactions between this motif in $GABA_BR1$ with the one in $GABA_BR2$ masks the presence of an endoplasmic reticulum retention sequence present in $GABA_BR1$ allowing the latter to be trafficked to the cell surface to form the functional $GABA_B$ receptor [59, 60]. In the case of δ opioid receptors studies show that although the C-terminal tail does not have a coiled-coil motif, it is thought to play a role in receptor dimerization since deletion of C-terminal 15 but not 7 amino acid residues leads to a significant decrease in the level of receptor homodimers [61].

An involvement of TM domains in GPCR dimerization has been suggested for a number of receptors including rhodopsin, β2-adrenergic receptors, dopamine receptors and opioid receptors [62-65]. In the case of rhodopsin, atomic force microscopy studies showed that it was arranged in dimeric arrays in native membranes obtained from wild-type mouse photoreceptors [62, 65]. In addition, molecular modeling studies indicated that TM4 and TM5 were involved in intradimeric contacts whereas interdimeric contacts probably involved TM1 and TM2 and the cytoplasmic loops connecting TM5 and TM6 [62]. In the case of β_2-adrenergic receptors, studies using a synthetic peptide corresponding to a glycophorin-like motif present in TM6 (^{272}LKTLGIIMGTFTL^{284}) showed that it interfered with the formation of receptor dimers; this implicated TM6 in β_2-adrenergic receptor dimerization [66]. Interestingly, a TM6 based synthetic peptide had no effect on D1 dopamine receptor dimer levels [66] suggesting that different TM interfaces could be involved in

GPCR dimerization. In the case of D2 dopamine receptors, although studies using synthetic peptides implicated the involvement of TM6 and TM7 in the dimerization of D2 dopamine receptors [67], substituted cysteine accessibility studies as well as cysteine cross-linking studies demonstrated an involvement of TM4 in the formation of the homodimer [68, 69].

In the case of opioid receptors a computational approach based on a combination of correlated mutation analysis and the structural information obtained from three dimensional models of the TM regions of the receptors built based on the crystal structure of rhodopsin was used to predict the TM interfaces involved in receptor homo/heterodimerization [70, 71]. These studies implicated TM4 and/or TM5 in δ opioid receptor dimerization, TM1 and TM3 in μ opioid receptor dimerization and TM5 in κ opioid receptor dimerization [71]. Moreover, this computational approach predicts the involvement of TM4, TM5 and TM6 of δ opioid receptors and TM1 of μ opioid receptor in μ-δ receptor heterodimerization [70]. In the case of δ opioid receptors, cysteine cross-linking studies support the involvement of TM4 and TM5 in receptor dimerization [72]. Finally, crystal structure analyses have implicated TM5 and TM6 in μ opioid and TM1 and TM2 in κ opioid receptor as dimer interfaces [45, 47]. Additional studies are needed to confirm the TM interfaces for μ and κ opioid receptor homodimerization as well as for μ-δ receptor heterodimerization.

CHAPTER 3

G Protein-Coupled Receptor Signaling

3.1 INTRODUCTION

The majority of neuropeptides exert their biological effects via activation of GPCRs. The binding of the neuropeptide to the extracellular region of the receptor leads to conformational changes in the TM region. These changes allow the intracellular domain of the receptor to activate the heterotrimeric G protein leading to intracellular signaling. As the name implies the heterotrimeric G protein has three subunits: α, β, and γ. The α subunit exchanges GDP for GTP upon activation. Activated G proteins are able to activate numerous downstream effector molecules (Fig. 3.1). The catalytic nature of the effectors leads to the generation of second messengers that, in turn, can activate enzymes resulting in signal amplification. This is a key feature of GPCR signaling. Another key feature is that of signal termination. The activated receptor, upon uncoupling from the G protein, undergoes phosphorylation that leads to the recruitment of adaptor proteins such as β-arrestin. This leads to receptor endocytosis and termination of the signal from the G protein cascade; this helps prevent possible harmful effects due to persistent receptor activation. A third key feature of GPCR signaling is that of signal diversification. This is achieved by the promiscuity of a specific GPCR to couple to more than one G protein, the ability of the different subunits of G proteins to activate multiple effectors, and the ability of a GPCR to signal via a non-G protein-mediated pathway. Over the last decade it has become increasingly evident that recruitment of β-arrestin to phosphorylated GPCRs can lead to activation of G protein independent signaling [73, 74]. In this chapter we will describe the signaling pathways activated following binding of an agonist to a GPCR as well as those involved in signal attenuation. We will also describe G protein independent signaling, the emerging concept of biased agonism, and how GPCR heteromerization expands the signaling repertoire of a receptor.

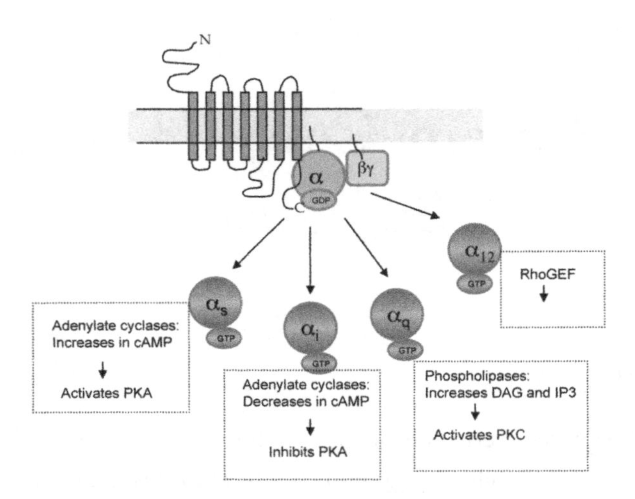

Figure 3.1. **Signaling cascades activated by GPCRs.** The Gα subunit associated with a GPCR can belong to either the Gαs, Gαi/o, Gαq/11 or Gα12/13 family. Activation of Gαs leads to activation of adenylyl cyclase and increases in cyclic AMP levels and in PKA activity. Activation of Gαi/o leads to inhibition of adenylyl cyclase and decreases in cyclic AMP levels and consequently in PKA activity. Activation of Gαq/11 leads to phospholipase-mediated increases in intracellular levels of DAG and IP3 and activation of PKC. Activation of Gα12/13 leads to activation of the small G protein Rho.

3.2 G PROTEIN-MEDIATED SIGNALING

In the basal state GPCRs are associated with heterotrimeric G proteins where the Gα subunit is bound to guanosine diphosphate (GDP), and in a complex with the Gβγ subunits. The binding of an agonist to a GPCR induces conformational changes to the receptor [10]. This, in turn, changes the conformation of receptor-associated heterotrimeric G proteins and induces an exchange of GDP for GTP at the Gα subunit. The GTP bound Gα subunit dissociates from the Gβγ complex leading to the activation of signaling cascades downstream of Gα and Gβγ and ultimately a cellular response [8–10, 75] (Fig. 3.2). The Gα subunit has intrinsic GTPase activity that promotes the hydrolysis of the bound GTP to GDP following which the Gα subunit re-associates with the Gβγ complex and the complex can now undergo another round of signaling events (Fig. 3.2). In addition, the duration of the activity of a G protein can be modulated by a number of proteins including regulators of G protein signaling (RGS proteins), activators of G protein signaling (AGS) and G protein receptor kinases.

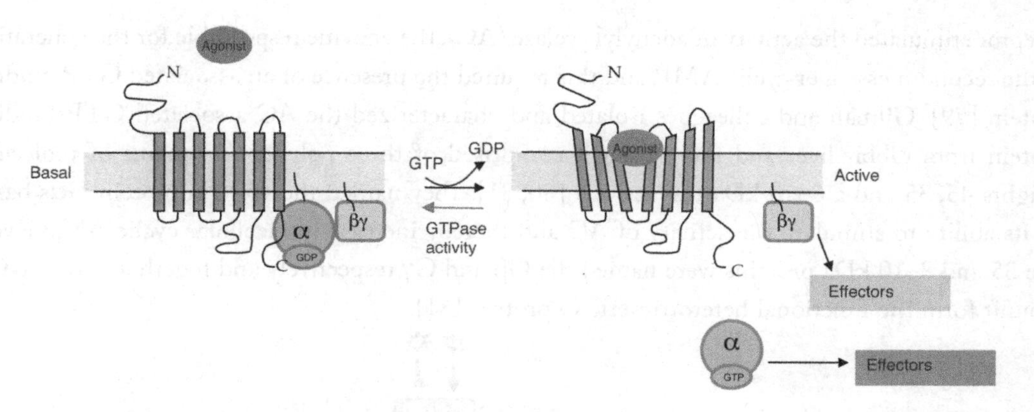

Figure 3.2. **Schematic representation of activation of heterotrimeric G proteins.** The binding of an agonist to a GPCR induces conformational changes that are transmitted to the receptor associated heterotrimeric G proteins. This leads to an exchange of the GDP bound to the Gα subunit (Gα-GDP) for GTP (Gα-GTP). The Gα-GTP dissociates from the Gβγ complex and both Gα-GTP and Gβγ activate downstream effectors. The intrinsic GTPase activity of the Gα subunit converts the bound GTP to GDP leading to the re-association of the Gα subunit with the Gβγ complex and another round of signaling events.

Different subtypes of Gα, Gβ and Gγ subunits have been identified and the signaling pathways activated by a given GPCR depend on which combination of heterotrimeric G proteins is associated with the receptor. For example, it has been shown that dopamine D1 receptors in medium spiny neurons in the striatum are associated with the Gαolf/β2/γ7 heterotrimer and signal through adenylyl cyclase V [76]. To date 20 different types of Gα subunits have been identified and they comprise four major Gα families: Gαs, Gαi/o, Gαq/11 and Gα12/13 [77]. Although different types of Gβ and Gγ subunits have been identified they are considered to form one functional unit, Gβγ, since they cannot be dissociated from each other except through the use of strong denaturing conditions. Classically, GPCRs are classified based on the Gα subunit associated with the receptor as Gαi/o, Gαs, Gαq/11, or Gα12/13 coupled receptors. The Gα subunits show some specificity for downstream effectors. For example Gαs family members activate and Gαi/o inhibit adenylyl cyclase activity, Gαq/11 family members activate phospholipase C activity while Gα12/13 modulate RhoGEF activity [77, 78]. The following sections describe the different Gα families with a focus on Gαs, Gαi/o and Gαq, the G-protein subunits found to be associated with neuropeptide receptors (see Chapter 5).

3.2.1 THE Gαs FAMILY

The involvement of G proteins in transmembrane receptor signaling was initially suggested by studies from Rodbell and colleagues that showed that the binding of the hormone glucagon to its receptor stimulated the activity of adenylyl cyclase (AC), the enzyme responsible for the generation of the second messenger cyclic AMP, and this required the presence of an associated GTP binding protein [79]. Gilman and colleagues isolated and characterized the AC associated GTP-binding protein from rabbit liver and found that it comprised of three polypeptide chains of molecular weights 45, 35 and a 8–10 kDa respectively [80, 81]. They named the 45 kDa protein Gαs based on its ability to stimulate the activity of AC and thereby increase intracellular cyclic AMP levels. The 35 and 8–10 kDa proteins were named the Gβ and Gγ respectively and together with the Gα subunit form the functional heterotrimeric G protein [81].

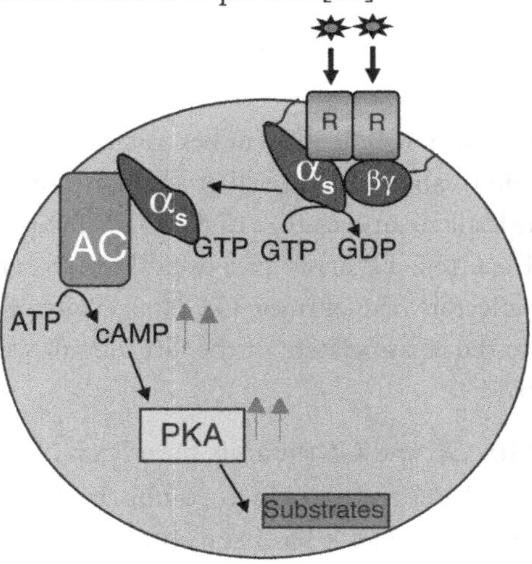

Figure 3.3. **Schematic of Gαs-mediated signaling.** Binding of an agonist (✸) to a Gαs-coupled receptor (R) leads to an exchange of GTP for GDP bound to Gαs. The Gαs-GTP activates adenylyl cyclase and consequently increases intracellular cyclic AMP levels. Increased levels of cyclic AMP are accompanied by an increase in the activity of PKA leading to increased phosphorylation of PKA substrates and ultimately a cellular response.

The Gαs family includes multiple isoforms of Gαs, that are the result of alternative splicing of exon 3, in addition to GαsXL, Gαolf and XL-Gαolf [82–87] (Table 3.1). Gαs isoforms are ubiquitously expressed and characterized by their sensitivity to cholera toxin obtained from the bacterium *Vibrio cholerae* (Table 3.1). The A1 subunit of cholera toxin catalyzes the covalent transfer of the ADP-ribose moiety from nicotinamide adesosine dinucleotide (NAD^+) to Arg201 on Gαs [88]. The ADP-ribosylated Gαs can bind GTP and activate AC. Under normal conditions activated Gαs (i.e. Gαs-GTP) is deactivated by its intrinsic GTPase activity. However, ADP-ribosylation inactivates the GTPase activity of Gαs rendering it constitutively active.

Table 3.1. Gα proteins and their effectors

Family	Family members	Encoding genes	Toxin sensitivity	Effectors	Tissue distribution	References
Gαs	7 isoforms produced by alternative splicing	*Gnas*	CTX	↑AC	Ubiquitous distribution; highest levels in hypothalamus, prefrontal cortex, pituitary and thyroid	[600, 601]
	GαsXL	*GnasXL*	CTX	↑AC	Developmentally regulated; present in midbrain, hindbrain & spinal cord mid-gestation; present in brain, pancreatic islets, pituitary & adrenal glands in adults	[600, 601]
	Gαolf	*Gnal*	CTX	↑AC	Olfactory epithelium; medium spiny neurons in striatum; cholinergic spiny interneurons	[600, 601]
	XLGαolf	*Gnal*	CTX	↑AC	Epigenetically regulated; hypothalamus; PFC; striatum	[600, 601]
Gαi/o	Gαi1	*Gnai1*	PTX	↓AC;↑ GIRK	Wide distribution; highest levels in amygdala, hypothalamus & spinal cord	[132, 133, 600]
	Gαi2	*Gnai2*	PTX	↓AC;↑ GIRK	Ubiquitous distribution; highest levels in immune cells	[132, 133, 600]
	Gαi3	*Gnai3*	PTX	↓AC;↑ GIRK	Wide distribution; highest levels in smooth muscle & immune cells	[132, 133, 600]
	Gαi2	*Gnai2*	PTX	↓AC;↑ GIRK	Ubiquitous distribution; highest levels in immune cells	[132, 133, 600]

	Gαi3	*Gnai3*	PTX	↓AC;↑ GIRK	Wide distribution; highest levels in smooth muscle & immune cells	[132, 133, 600]
	Gαo	*Gnao*	PTX	↓VDCC;↑ GIRK	Neuroendocrine distribution; heart	[128, 129, 135, 600]
	Gαz	*Gnaz*	PTX insensitive	↓AC;↑ GIRK	Neuronal; platelets	[150–153, 600, 602]
	Gαt-r	*Gnat1*	PTX	↑cGMP -PDE	Retinal rods; taste cells	[600]
	Gαt-c	*Gnat2*	PTX	↑cGMP -PDE	Retinal cones; taste cells	[600]
	Gαgust	*Gnat3*	PTX	↑PDE6	Taste cells	[600]
Gαq/11	Gαq	*Gnaq*		↑PLCβ	Ubiquitous distribution; highest levels in prefrontal cortex & amygdala	[161–163, 600]
	Gα11	*Gna11*		↑PLCβ	Almost ubiquitous distribution; highest levels in lungs, prostate, testis, cerebellum & prefrontal cortex	[161–163, 600]
	Gα14	*Gna14*		↑PLCβ	Kidneys; lungs; spleen; testis	[161-164, 600]
	Gα15/16	*Gna15*		↑PLCβ	Hematopoietic cells	[161-165, 600]
Gα12/13	Gα12	*Gna12*		RhoGEF; Btk GAP 1^{m}; cadherin	Ubiquitous distribution; highest levels in lungs, prostate, smooth muscle, hypothalamus & spinal cord	[600]
	Gα13	*G α13*		RhoGEF; radixin; cadherin	Ubiquitous distribution; highest levels in immune cells	[600]
AC, adenylyl cyclase; cGMP, cyclic guanosine monophosphate; GIRK, G protein-gated inwardly rectifying K^+ channels; PDE, phosphodiesterase; PLC, phospholipase C; RhoGEF, Rho guanine exchange factor; VDCC, voltage-dependent Ca^{+2} channels						

Gαs is anchored to the plasma membrane by palmitoylation of its amino terminal region [89-91]. Activation of the Gαs subunit leads to depalmitoylation and release of Gαs from Gβγ and the plasma membrane. The released Gαs subunit can now directly activate AC and other effectors [92–95] (Fig. 3.3). Studies indicate that the GTPase activity of the Gαs subunit may be influenced by RGS proteins and by AC itself [96, 97].

In addition to participating in receptor mediated signaling, Gαs may also play a role in regulating endocytic vesicular trafficking although the mechanisms are not clearly understood [96, 98, 99]. Very little information is available about the mechanism of degradation of Gαs. Studies show that following receptor activation Gαs internalizes into vesicles that show minimal overlap with receptor-containing vesicles [100] indicating that receptors and G proteins are differentially sequestered. A recent study showed that Gαs could be ubiquitinated and possibly degraded through the proteasome [101]. However, further studies are needed to elucidate the precise mechanisms of degradation of Gαs.

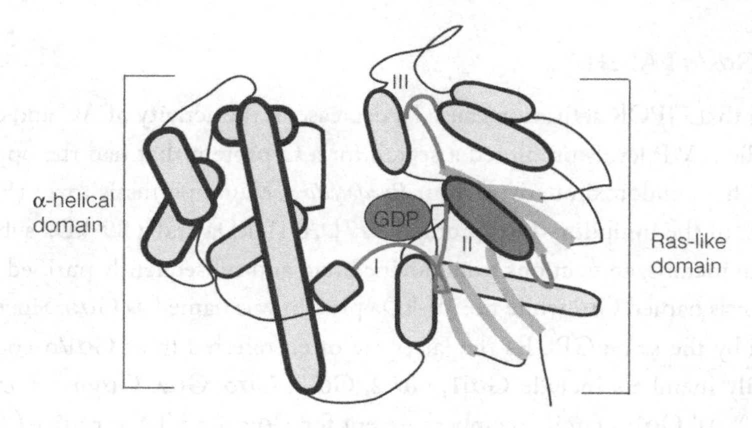

Figure 3.4. **Gαs structure showing switches I, II and III in red,** ▭α-helix; ■ β-sheet; ■ switches I, II and III.

The crystal structure of the Gαs isoform shows that it has two structural domains: a Ras-like GTPase domain and a helical domain (Fig. 3.4). The arrangement of the GTPase and helical domains forms a deep cleft to which guanine nucleotides bind [102, 103]. The Ras-like GTPase domain plays a role in the hydrolysis of GTP and is involved in binding to the receptor, Gβγ, and effector proteins [104–106]. The helical domain is made up of six α-helical bundles that form a lid over the nucleotide binding pocket that is thought to prevent the release of bound GDP in the inactive state [104–106]. Receptor activation leads to an exchange of GDP for GTP at the Gα subunit. This process involves conformational changes of three flexible loops present in the GTPase domain that are known as switches I (residues 173–183), II (residues 195–215) and III (residues 227–238); these conformational changes facilitate the binding of GTP and the dissociation of the Gβγ subunit [106–109] (Fig. 3.4). Switch I forms one of the two strands connecting the GTPase and the α-helical domains. Switch II assumes a partial helical conformation in the active state. This region is involved in interactions between Gα-GDP and Gβγ and between Gα-GTP and downstream effectors, RGS proteins, or GoLoco motifs. Switch III assumes an ordered loop structure only in the active conformation of the Gα protein [109–112]. Although conformational changes in switches I and II are influenced by the presence of the γ-phosphate group of GTP, switch III has no direct contact with the bound nucleotide [113]. However, it is thought that the stabilization of the active conformation of Gαs as well as high affinity GTP binding involves interactions between switches II and III as well as between switch III and the helical domain [103, 113–118] while stabilization of the inactive conformation involves interactions between the amino terminus of Gαs, switch II and the Gβ subunit [109, 119]. In addition, it has been proposed that interactions between the receptor and G proteins occur primarily via the carboxyl terminal of Gαs [120–126].

3.2.2 THE Gαi/o FAMILY

Studies showing that GPCR activation caused a decrease in the activity of AC and consequently of intracellular cyclic AMP levels prompted a search for a G protein that had the opposite effects of Gαs. Studies with an endotoxin derived from *Bordetella pertussis*, pertussis toxin (PTX), helped in the identification of the inhibitory Gαi protein [127]. A 41-kDa and a 39-kDa substrate for PTX were identified in membrane fractions from bovine brain and subsequently purified [128, 129]. The 41-kDa protein was named Gαi while the 39-kDa protein was named as Gαo. Since Gαo and Gαi can be activated by the same GPCRs the latter are often referred to as Gαi/o-coupled receptors. The Gαi/o family members include Gαi1, Gαi2, Gαi3, Gαo, Gαz, Gαgust, Gαt-r, and Gαt-c [130] (Table 3.1). All Gαi/o family members, except for Gαz, are PTX sensitive (Table 3.1). PTX is an ADP-ribosyl-transferase that catalyzes the addition of ADP-ribose to a Cys residue present at the fourth position from the C-terminus of the Gαi/o subunit. This is an irreversible modification that prevents interactions between the receptor and the heterotrimeric G protein. The Gαi/o subunit remains in the GDP-bound inactive state and cannot inhibit adenylyl cyclase activity or open K^+ channels [130]. The characteristics of the Gαi, Gαo and Gαz proteins are described below.

Gαi

Gαi proteins are structurally very similar to Gαs [131]. The three Gαi isoforms, Gαi1, Gαi2, and Gαi3, are the products of three different genes [132] and exhibit ~85% sequence homology and partially overlapping expression patterns with Gαi1 being primarily expressed in the nervous system, Gαi2 being ubiquitously expressed while Gαi3 is broadly expressed in peripheral tissues [133] (Table 3.1). Of the three Gαi isoforms, Gαi2 is the most predominant [133]. In contrast to Gαs proteins that are palmitoylated at the amino terminal region, Gαi subunits are both myristoylated and palmitoylated [134]. The interaction of Gαi with its GPCR induces an exchange of GDP for GTP at the Gαi subunit and this activation causes an inhibition of AC activity [131] (Fig. 3.5). Studies show that in addition to AC, Gαi can interact with other effectors such as those that regulate phosphatidylinositol turnover, the activation of potassium channels and inhibition of voltage-dependent Ca^{2+} channels [131] (Table 3.1).

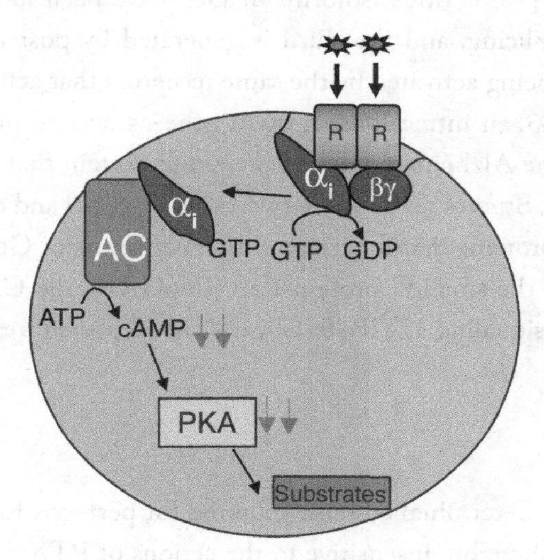

Figure 3.5. **Schematic of Gαi-mediated signaling.** Binding of an agonist (✸) to a Gαi-coupled receptor (R) leads to an exchange of GTP for GDP bound to Gαi. The Gαi-GTP inhibits adenylyl cyclase and consequently decreases intracellular cyclic AMP levels. Decreased levels of cyclic AMP are accompanied by a decrease in the activity of PKA leading to decreased phosphoryla-tion of PKA substrates and ultimately a cellular response.

The crystal structure of one member of this family, Gαi1, has been reported and it shows that like Gαs it has two structural domains: a Ras-like GTPase domain and an helical domain [102]. Comparison of the structure of Gαi1-GDP to that of Gαi1-GTP shows that in the GDP bound inactive state the N- and C-terminal regions of Gαi1, that are involved in binding to the receptor and to Gβγ subunits, are ordered into compact microdomains [102]. Following binding of GTP to Gαi1 these regions become disordered [102]. Moreover, GTP binding induces conformational changes in switch I (residues 177–187), II (residues 199–219), III (residues 231–242) and IV regions (residues 111–119) as well as a movement of the linker region connecting the GTPase and helical domains that is thought to facilitate the release of GDP [102]. Switch IV is thought to be involved in the formation of Gαi1 oligomers in the basal state [102]. The stabilization of the active conformation of Gαi appears to involve interactions between switches I and II and the γ-phosphate of bound GTP and between switches II, III and the α-helical domain [102].

$G\alpha_o$

$G\alpha_o$ is the most abundant G protein present in the mammalian brain thought to constitute ~2% of membrane protein in neurons [128, 129]. Outside the brain it has a more restricted distribution being present only in endocrine cells and heart and at much lower levels compared to other heterotrimeric G proteins [135]. Three isoforms of $G\alpha_o$ have been identified, two of which are generated by alternative splicing, and the third is generated by post-translational modification [136–138]. In addition to being activated by the same receptors that activate Gαi [139–144], $G\alpha_o$ can be activated by GAP43, an intracellular growth cone-associated protein involved in neurite outgrowth [145], and by the Alzheimer amyloid precursor protein that is responsible for familial forms of this disease [146]. Studies using yeast-two hybrid screens and cDNA expression cloning have identified candidate proteins that function as direct effectors of $G\alpha_o$. These include the GT-Pase activating protein for the small G protein Rap (RapGAP), the GAP for $G\alpha_z$ ($G\alpha_z$-GAP), the regulator of G protein signaling 17 (RGS-17), and the G protein regulated inducer of neurite outgrowth (GRIN) [147-149].

Gαz

The Gαz protein lacks the C-terminal cysteine required for pertussis toxin-catalyzed ADP-ribosylation [150, 151] and is therefore insensitive to the actions of PTX; hence it is called PTX-insensitive G protein. Gαz is expressed predominantly in neuronal tissues and cells [150, 151] (Table 3.1). Like Gαi, Gαz can inhibit the activity of AC and stimulate that of K^+ channels [152, 153]. Interestingly, Gαz has unique properties that distinguish it from other Gα proteins: (i) covalent modifications of Gαz modulate its signaling [154–156]; (ii) the kinetics of GTP hydrolysis is much slower than other Gα proteins [157]; (iii) phosphorylation by protein kinase C (PKC) or by p21-activated kinase (PAK1) decreases its affinity for the Gβγ complex and keeps the Gαz protein active for a longer period of time [158, 159]; and (iv) Gαz interacts with several RGS proteins [157, 160].

3.2.3 THE Gαq/11 FAMILY

Gαq/11 family of G proteins differs from Gαs and Gαi in that phospholipase Cβ (PLC-β) rather than AC is the main effector of its activity [161–163]. Members of this family include Gαq, Gα11, Gα14 and the mouse Gα15/human Gα16 homologs (Table 3.1). Gαq and Gα11 share 88% amino acid homology, are widely distributed in mammalian tissues and can activate different isoforms of PLC-β (i.e. PLC-β1 to PLC-β4) [163]. Gα14 exhibits 81% sequence homology to Gαq and is expressed in the spleen, lung, kidney, and testis [164]. The human Gα16 and the mouse Gα15 homolog share 57% homology with Gαq and are expressed in hematopoietic cells [164, 165].

Activation of PLC-β by Gαq/11 family members induces the hydrolysis of phosphatidylinositol 4,5-bisphosphate (PIP_2) to produce inositol triphosphate (IP3) and diacylglycerol (DAG) leading to the activation of downstream signaling (Fig. 3.6). DAG remains bound to the membrane and acts as a second messenger that activates PKC. IP3 is released as a soluble moiety into the cytosol and diffuses through the cytosol and binds to IP3 receptors such as calcium channels in the endoplasmic reticulum (ER). These channels are specific to Ca^{+2} allowing its passage into the cytoplasm, thereby increasing intracellular Ca^{+2} levels and leading to the activation of a cascade of intracellular Ca^{+2}-dependent signaling events [161, 166, 167] (Fig. 3.6).

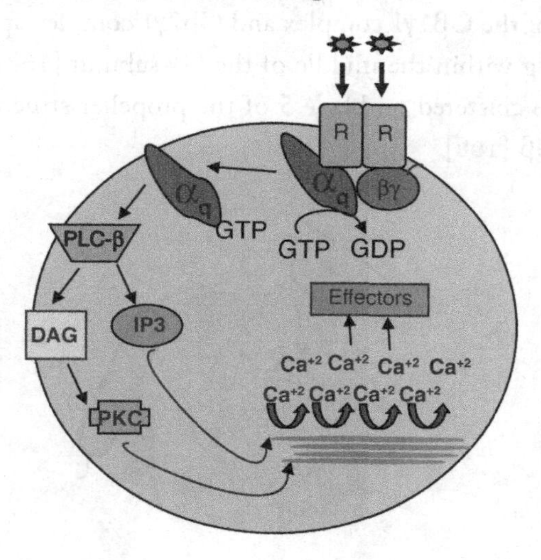

Figure 3.6. **Schematic of Gαq-mediated signaling.** Binding of an agonist (✸) to a Gαq-coupled receptor (R) leads to an exchange of GTP for GDP bound to Gαq. The Gαq-GTP activates phospholipase Cβ leading to an increase in the intracellular levels of DAG and IP3. DAG activates PKC leading to the phosphorylation of PKC substrates. IP3 binds to IP3 receptors in the ER which leads to a release of Ca^{+2} from intracellular stores and activation of downstream signaling cascades leading to a cellular response.

3.2.4 THE Gβγ FAMILY

The Gβ and Gγ subunits associate with each other to form one functional unit, Gβγ. A number of Gβ isoforms (Gβ1-5) encoded by different genes have been identified [168–170] (Table 3.2). The Gβ1–4 subunits share 78–88% sequence homology while Gβ5 shares only 51–53% homology with the other Gβ subunits [170, 171]. Gβ1-Gβ4 exhibit a broad tissue distribution while Gβ5 is primarily expressed in the central nervous system [169].

The Gβ subunit is characterized by a repeating sequence motif called the WD repeat and the presence of an amino terminal α-helix of ~20 amino acids. The WD motif is characteristic of proteins that form large macromolecular assemblies [168]. The crystal structure of the heterotrimeric G proteins shows that the WD repeat in Gβ is made up of β strands arranged in a ring forming a seven bladed propeller structure where each blade is made up of four twisted β strands (Fig. 3.7). The ring shape of the propeller structure is maintained by a β strand coming from the N-terminal region of Gβ and three beta strands from the C-terminal region [168, 172–174] (Fig. 3.7). Residues that determine the specificity of Gβ and Gγ interactions have been identified. For example, the specificity of the assembly of the Gβ1γ1 complex and Gβ2γ1 complex appears to be determined by a stretch of 14 residues lying within the middle of the Gγ subunit [168]. In addition, the residues in Gβ that contact Gγ1 are clustered on blade 5 of the propeller structure and a small section of the N-terminal region of Gβ [168].

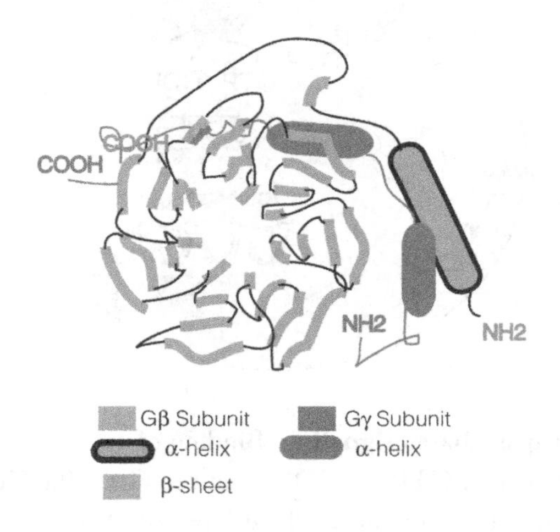

Figure 3.7. **Schematic representation of the structure of Gβγ subunits.** The Gβ subunit (blue) is characterized by the presence of a N-terminal α-helical domain and a WD repeat motif made up of β-strands arranged in a ring to form a seven bladed propeller structure. The Gγ subunit (red) comprises two α-helical domains joined by an intervening loop.

Over thirteen different Gγ subunits subunits have been identified that are encoded by different genes [170] (Table 3.2). The Gγ subunits can be modified by isoprenylation, methylation, farnesylation, or geranylgeranylation and this helps anchor this subunit to the plasma membrane [168–170]. The Gγ subunit comprises an extended stretch of two α-helices joined by an intervening loop. It has been proposed that the N-terminal region of Gγ forms a coiled-coil interaction with the N-terminal α-helix of the Gβ subunit while the rest of the Gγ subunit binds to the outer edge of the Gβ subunit [119, 175].

Emerging studies have shown that the Gβγ complex can activate a number of downstream effectors. The first effector of the Gβγ complex to be discovered was the G protein-gated inwardly rectifying potassium channel (GIRK) [176]. Since then, a number of downstream effectors of Gβγ have been identified including AC, phospholipases Cβ1 to 3, and PI3K among others [168, 170] (Table 3.2).

Table 3.2. Gβγ proteins and their effectors

Family	Family members	Encoding genes	Effectors	Tissue distribution	References
Gβ	Gβ1	*Gnb1*	↓AC1;↑AC2,4 & 7;↑PLCβ; ↑GIRK1-4;↑PLA2; ↑GRK2 &3;↑PI-3-K β & γ;↓VDCC	Wide distribution; highest levels in brain & hematopoietic cells	[168–170, 176, 600]
	Gβ2	*Gnb2*	↓AC1;↑AC2,4 & 7;↑PLCβ; ↑GIRK1-4;↑PLA2; ↑GRK2 &3;↑PI-3-K β & γ;↓VDCC	Wide distribution; highest levels in blood and lungs	[168–170, 176, 600]
	Gβ3	*Gnb3*	↓AC1;↑AC2,4 & 7;↑PLCβ; ↑GIRK1-4;↑PLA2; ↑GRK2 &3;↑PI-3-K β & γ;↓VDCC	Wide distribution; retinal cones; highest levels in cortex & pituitary	[168–170, 176, 600]
	Gβ4	*Gnb4*	↓AC1;↑AC2,4 & 7;↑PLCβ; ↑GIRK1-4;↑PLA2; ↑GRK2 &3;↑PI-3-K β & γ;↓VDCC	Wide distribution; highest levels in lungs & placenta	[168–170, 176, 600, 603]
	Gβ5	*Gnb5*	↓AC1;↑AC2,4 & 7;↑PLCβ; ↑GIRK1-4;↑PLA2; ↑GRK2 &3;↑PI-3-K β & γ;↓VDCC	Mainly brain	[168–170, 176, 600]
Gγ	Gγ; Gγrod	*Gngt1*	↓AC1;↑AC2,4 & 7;↑PLCβ; ↑GIRK1-4;↑PLA2; ↑GRK2 &3;↑PI-3-K β & γ;↓VDCC	Retinal rods; brain	[168, 170, 176, 600]

	Gγ14; Gγcone	*Gngt2*	↓AC1;↑AC2,4 & 7;↑PLCβ; ↑GIRK1-4;↑PLA2; ↑GRK2 &3;↑PI-3-K β & γ;↓VDCC	Retinal cones; brain	[168, 170, 176, 600]
	Gγ2; Gγ6	*Gng2*	↓AC1;↑AC2,4 & 7;↑PLCβ; ↑GIRK1-4;↑PLA2; ↑GRK2 &3;↑PI-3-K β & γ;↓VDCC	Wide distribution; highest levels in lungs & placenta	[168, 170, 176, 600]
	Gγ3	*Gng3*	↓AC1;↑AC2,4 & 7;↑PLCβ; ↑GIRK1-4;↑PLA2; ↑GRK2 &3;↑PI-3-K β & γ;↓VDCC	Brain; blood	[168, 170, 176, 600]
	Gγ4	*Gng4*	↓AC1;↑AC2,4 & 7;↑PLCβ; ↑GIRK1-4;↑PLA2; ↑GRK2 &3;↑PI-3-K β & γ;↓VDCC	Brain; other tissues	[168, 170, 176, 600]
	Gγ5	*Gng5*	↓AC1;↑AC2,4 & 7;↑PLCβ; ↑GIRK1-4;↑PLA2; ↑GRK2 &3;↑PI-3-K β & γ;↓VDCC	Wide distribution; highest levels in heart, lungs & hematopoietic cells	[168, 170, 176, 600]
	Gγ7	*Gng7*	↓AC1;↑AC2,4 & 7;↑PLCβ; ↑GIRK1-4;↑PLA2; ↑GRK2 &3;↑PI-3-K β & γ;↓VDCC	Wide distribution; highest levels in hematopoietic cells	[168, 170, 176, 600]
	Gγ8; Gγ9	*Gng8*	↓AC1;↑AC2,4 & 7;↑PLCβ; ↑GIRK1-4;↑PLA2; ↑GRK2 &3;↑PI-3-K β & γ;↓VDCC	Olfactory epithelium	[168, 170, 176, 600]
	Gγ10	*Gng10*	↓AC1;↑AC2,4 & 7;↑PLCβ; ↑GIRK1-4;↑PLA2; ↑GRK2 &3;↑PI-3-K β & γ;↓VDCC	Wide distribution; highest levels in brain, lungs & hematopoietic cells	[168, 170, 176, 600, 604]

	Gγ11	*Gng11*	↓AC1;↑AC2,4 & 7;↑PLCβ; ↑GIRK1-4;↑PLA2; ↑GRK2 &3;↑PI-3-K β & γ;↓VDCC	Wide distribution; highest levels in lungs, adipose tissue & platelets	[168, 170, 176, 600, 604]
	Gγ12	*Gng12*	↓AC1;↑AC2,4 & 7;↑PLCβ; ↑GIRK1-4;↑PLA2; ↑GRK2 &3;↑PI-3-K β & γ;↓VDCC	Wide distribution; highest levels in thyroid; placenta & smooth muscle	[168, 170, 176, 600]
	Gγ13	*Gng13*	↓AC1;↑AC2,4 & 7;↑PLCβ; ↑GIRK1-4;↑PLA2; ↑GRK2 &3;↑PI-3-K β & γ;↓VDCC	Taste buds; brain	[168, 170, 176, 600]

AC, adenylyl cylcase; cGMP, cyclic guanosine monophosphate; GIRK, G protein-gated inwardly rectifying K^+ channels; GRK, G protein-coupled receptor kinase; PDE, phosphodiesterase; PI-3-K, phosphoinositide-3-kinase; PLA, phospholipase A; PLC, phospholipase C; RhoGEF, Rho guanine exchange factor; VDCC, voltage-dependent Ca^{+2} channels

3.3 DIRECT EFFECTORS OF G PROTEIN-MEDIATED SIGNALING

3.3.1 ADENYLYL CYCLASES

Adenylyl cyclases are direct effectors of Gαs and Gαi/o mediated signaling. Adenylyl cyclases are enzymes that catalyze the cyclization of ATP to generate cyclic AMP and inorganic pyrophosphate. To date at least nine closely related AC isoforms (AC1–AC9) and two splice variants of AC8 have been identified that are obligatory membrane proteins [177–181] (Table 3.3). ACs are grouped into four subfamilies, Group 1 (AC1, AC3, AC8), Group 2 (AC2, AC4, AC7), Group 3 (AC5 and AC6) and Group 4 (AC9) based on sequence similarity and regulatory properties [182]. In the last decade soluble adenylyl cyclases have been identified; however, they are insensitive to heterotrimeric G proteins and are regulated by bicarbonate and calcium [183–186].

Each AC isoform has a specific pattern of tissue/organ distribution and a specific pattern of regulation by G proteins [187–189] (Table 3.3). The most abundant isoforms expressed in the mammalian brain are AC1, AC2 and AC9. The mRNA levels for AC1 and AC2 are high in hippocampus, cerebral cortex and cerebellum [179]. In addition to the brain, the AC isoforms, AC4-7 and AC9, can also be detected to different extents in peripheral tissues including heart, kidney, liver and muscle [190, 191]. Studies show that the activity of all membrane bound isoforms of AC is stimulated by Gαs proteins [192, 193], and that all AC isoforms except for AC9 are stimulated by forskolin (a diterpene derived from geranylgeranyl pyrophosphate) [194, 195]. The activities of AC1, AC3, AC8 can also be stimulated by Ca^{2+}-calmodulin [182, 196–198], of AC2, AC4, AC7 by Gβγ [199–202], and of AC2, AC5 and AC7 by PKC [190, 191] (Table 3.3). Interestingly, the activities of AC1, AC5, and AC6 can be inhibited by Gαi proteins, of AC1 and AC5 by Gβγ, of AC1 by CAMK IV, AC3 by CAMK II, AC5 and AC6 by PKA and by Ca^{2+} in a calmodulin independent manner, of AC6 by PKC and of AC9 by calcineurin [190, 191] (Table 3.3). All membrane bound AC isoforms are inhibited via a dead-end non-competitive mechanism by P-site inhibitors which are adenine nucleoside -3'-polyphosphates [203, 204] (Table 3.3).

Table 3.3. Characteristics of adenylyl cyclase isoforms

Isoforms	Encoding gene	Tissue distribution	Activators	Inhibitors	Phenotype of knock-out mice	References
AC1	*ADCY1*	Brain, retina, spinal cord; adrenal gland	Gαs; forskolin; Ca^{+2}-calmodulin	Gαi; Gβγ; CAMK IV; P-site analogs	Normal acute pain responses; ↓ responses in inflammatory and neuropathic pain models; ↑ ethanol-induced neurodegeneration; ↓ opioid dependence	[178, 190, 191, 605–610]
AC2	*ADCY2*	Brain; olfactory bulb; spinal cord; heart; lung; muscle	Gαs; forskolin; Gβγ; PKC	P-site analogs		[178, 190, 191, 609, 610]
AC3	*ADCY3*	Olfactory neurons; brain; heart; lungs; testes; brown tissue	Gαs; forskolin; Ca^{+2}-calmodulin	CAMK II; P-site analogs	Peripheral & behavioral anosmia	[178, 190, 191, 609–611]
AC4	*ADCY4*	Kidney; brain; liver; spinal cord; heart; lungs; testes; brown tissue; adrenal gland	Gαs; forskolin; Gβγ	P-site analogs	No obvious phenotype	[178, 190, 191, 609, 610]

AC5	*ADCY5*	Brain; heart; spinal cord; lungs; testes; brown tissue; kidney; liver; adrenal gland	Gαs; forskolin; PKC and ζ	Gαi; Gβγ; PKA; Ca^{+2}; P-site analogs	Protection against pressure-induced myocyte apoptosis; ↓ dopamine D2 receptor function; loss of behavioral effects of selective μ & δ but not κ agonists	[178, 190, 191, 609, 610, 612–614]
AC6	*ADCY6*	Heart; brain; kidney; testes; liver; spinal cord; lungs; brown tissue; muscle; adrenal gland	Gαs; forskolin	Gαi; PKA;PKC; Ca^{+2}; P-site analogs	Cardiac AC6 knock-out mice exhibit ↓left ventricular contraction; ↓in β-adrenergic-mediated cAMP levels in cardiac myocytes	[178, 190, 191, 609, 610, 615]
AC7	*ADCY7*	Lungs; heart; spleen; kidney; brain; testes; liver; muscle	Gαs; forskolin; Gβγ; PKC	P-site analogs	Pre- & post-natally lethal; ↓ depression-like behavior in heterozygous females	[178, 190, 191, 609, 610]
AC8	*ADCY8*	Brain; spinal cord	Gαs; forskolin	P-site analogs	↓responses in inflammatory and neuropathic pain models;↓opioid dependence; ↑ ethanol-induced neurodegeneration; ↑ acquisition of spatial information	[178, 190, 191, 605–610, 616]
AC9	*ADCY9*	Skeletal muscle; brain; lung; liver; spinal cord; heart; kidney; adrenal gland	Gαs	Calcineurin; P-site analogs	Altered immune system; IgG1 response to ovalbumin challenge	[178, 190, 191, 609, 610]
CAMK, Ca^{+2}/calmodulin dependent kinase; PKA, protein kinase A; PKC, protein kinase C						

The different isoforms of AC share the same predicted three-dimensional structure. AC has two hydrophobic domains, each spanning the membrane six times (Fig. 3.8). The cytoplasmic domains in the C1 and C2 region can be subdivided into a and b domains; of these the C1a and C2a domains interact to form the catalytic site of the enzyme that binds to ATP to convert it to cyclic AMP. The C1b region is present in all AC isoforms while the C2b region is present only in AC1, AC2, AC3 and AC8 isoforms [205]. The crystal structure of the soluble C2a domain of AC2 in complex with the C1a domain of AC5 in the presence of forskolin and GTPγS-bound Gαs has been obtained [105] and shows that the switch II and the α3-β5 loop of Gαs makes contact with AC [105].

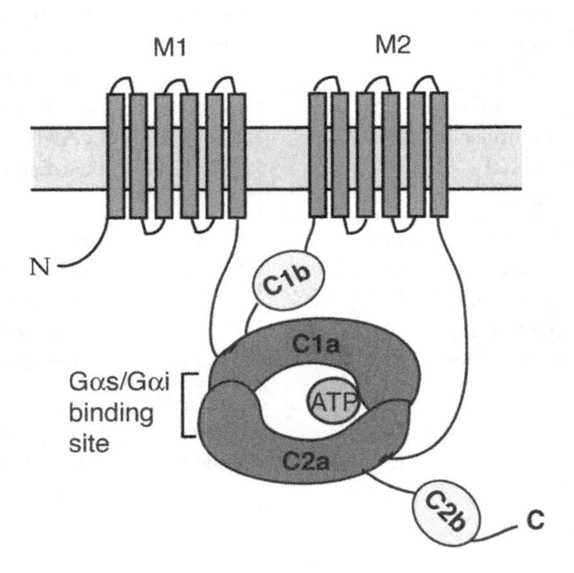

Figure 3.8. **Schematic representation of adenylyl cyclase structure.** Adenylyl cyclase is a 12 membrane spanning protein that is organized into (i) two hydrophobic domains (M1 & M2) that each traverse the membrane six times; (ii) a cytoplasmic catalytic site comprising the C1a & C2a domains that is regulated by Gαs/Gαi; and (iii) a cytoplasmic non-catalytic site comprising the C1b & C2b domains.

3.3.2 PHOSPHOLIPASES

Phospholipases of family C (PLC) are direct effectors of Gαq/11 and Gαi/o-mediated signaling. Gαq/11 family members can activate PLC β1-4 isoforms [206] while Gαi/o members activate PLCε [207]. The activity of PLCε can also be stimulated by constitutively active Gα12 as well as various Gβγ dimers [207–209]. Phospholipase C is an enzyme that catalyzes the hydrolysis of the membrane phospholipid, phosphatidylinositol-4,5-bisphosphate (PIP2) to generate two intracellular messengers, DAG and inositol-1,4,5-triphosphate (IP3) that are involved in the activation of PKC and intracellular Ca^{2+} mobilization [210]. To date, thirteen mammalian PLC isozymes have been identified that are encoded by different genes in humans (Table 3.4). They are classified as β, γ, δ, ε, ζ, and η based on their primary structure [211] (Table 3.4).

Table 3.4. Characteristics of PLC isoforms

Isoforms	Encoding gene	Tissue distribution	Regulators	Phenotype of knock-out mice	References
PLC-β1	*PLCB1*	Brain; adrenal gland; lungs	Gαq; PI^3P	Induces epilepsy	[166, 211, 216, 617]
PLC-β2	*PLCB2*	Brain; hematopoietic cells	Gβγ; Gαq	↓chemoattractant-mediated Ca^{+2} release & superoxide production in neutrophils; ↓in acid, sweet & bitter taste perception	[166, 211, 216, 618–620]
PLC-β3	*PLCB3*	Brain; liver; parotid gland; platelets	Gβγ; Gαq; PDZ domain	↓morphine-mediated antinociception; ↓agonist-mediated Ca^{+2} current reduction; ↑myeloproliferative disease, lymphomas & other tumors; ↓atherosclerosis	[166, 211, 216, 621–623]
PLC-β4	*PLCB4*	Brain; cerebellum; retina	Gαq	↓ visual processing abilities; ataxia; ↓induction of mGluR1-mediated long term depression	[166, 211, 216, 617, 624, 625]
PLC-γ1	*PLCG1*	Brain; lungs; thymus	Tyrosine kinase; PI_3P	Empryonically lethal	[211, 222, 223]
PLC-γ2	*PLCG2*	Lungs; thymus; spleen	Tyrosine kinase; PI_3P	Protection from clinical signs of arthritis & from arthritis induced loss of articular function; ↓ mature B cells	[211, 224, 225]
PLC-δ1	*PLCD1*	Brain; lungs; heart; testis; spleen; skeletal muscle	Ca^{+2}; transglutaminase II (Gαh)	↑hair loss; epidermal hyperplasia; ↑spontaneous skin tumors; PLC-δ1/PLC-δ3 double knock-out mice are embryonically lethal	[166, 211, 216, 237, 626, 627]

PLC-δ3	*PLCD3*	Brain; heart; skeletal muscle	Ca^{+2}	PLC-δ1/PLC-δ3 double knock-out mice are embryonically lethal with defects in placental development	[166, 211, 216, 237]
PLC-δ4	*PLCD4*	Skeletal muscle; testis; sperm	Ca^{+2}	Sperm unable to initiate acrosome reaction	[166, 211, 216, 237, 628, 629]
PLC-ε	*PLCE1*	Skeletal muscle; lungs; heart; liver	Ras; Gβγ; Gα12	↓chemical carcinogen skin tumor development; congenital heart malformations	[166, 211, 233, 630–632]
PLC-ζ	*PLCZ1*	Sperm	Ca^{+2}		[211, 235]
PLC-η1	*PLCL1*	Brain; spinal cord	Ca^{+2}		[166, 211, 236, 237, 633]
PLC-η2	*PLCL2*	Brain; retina	Ca^{+2}; Gβγ	No phenotype	[166, 211, 237, 237, 633]

The different PLC isoforms share a conserved core architecture comprising an N-terminal pleckstrin homology (PH) domain (except for PLCζ) followed by a series of four EF-hand domains (helix-turn-helix structural domains that bind Ca^{+2} ions), a catalytic TIM barrel comprised of two halves (X and Y boxes) and a C-terminal C2 domain [166, 167] (Fig. 3.9). The catalytic TIM barrel domain is the most conserved domain among the PLC isoforms [166, 167]. The X and Y boxes are separated by a linker region and fold into an alternative pattern having α helices on the outside and β strands on the inside of the TIM barrel to form the active site of PLC [212, 213]. Studies suggest that the X/Y linker may play a role in the auto-inhibition of PLC isozyme activity [167]. The C2 domain forms an eight β strand antiparallel sandwich, where three loops at one end of the sandwich form Ca^{+2} binding sites [214]. Although all PLC isozymes have the same enzymatic activity and share common structural domains they are differentially regulated [167, 215]. This could be due to the presence of isozyme specific regulatory domains. The different PLC isoforms are described below.

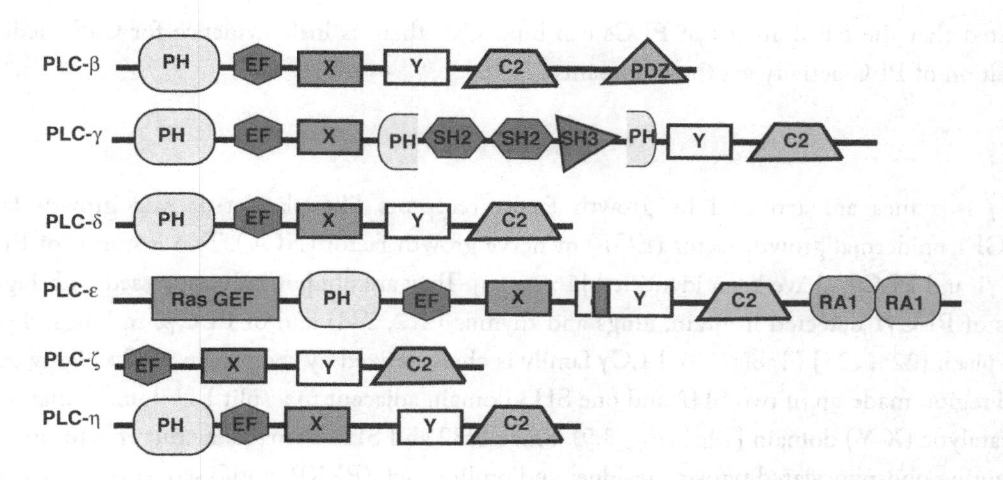

Figure 3.9. **Structural organization of PLC isoforms.** The core architecture of PLC isoforms comprise of a pleckstrin homology domain (PH), four helix-turn-helix structural domains that bind Ca^{+2} ions (EF), a catalytic domain (X and Y boxes) and a C2 domain. Additional domains present depending on the PLC isoform include a PDZ binding motif (PDZ), a Src homology domain (SH), a RasGTPase exchange factor-like domain (Ras GEF) or a Ras associating domain (RA).

PLCβ

PLCβ is a direct effector of Gαq/11-coupled GPCRs [206]. Four isoforms of PLCβ, PLCβ1-PLBβ4, have been identified. PLCβ1 is widely expressed particularly in the brain where it is found at high levels in the cerebral cortex and hippocampus (Table 3.4). PLCβ2 is expressed mainly in hematopoietic cells and in the brain (Table 3.4). PLCβ3 is expressed in liver, parotid gland, platelets and at low levels in the brain (Table 3.4). PLCβ4 is expressed in the brain with high expression levels in the cerebellum and retina [166, 216] (Table 3.4).

The PLCβ family is characterized by the presence of an unique C-terminal coil-coiled "PDZ" domain that is thought to be important for dimerization, membrane association and activation by Gα subunits [167]. PLCβ family members are differentially regulated by Gαq and Gβγ subunits. Gαq has a higher affinity for PLCβ1 and PLCβ3 than for PLCβ2 [217]. In addition, all four PLC isoforms can be activated Gβγ [217]. The N- and C-terminal ends of the C2 domain provide the major binding surface for activated Gαq while Gβγ has been shown to bind to the PH domain of PLCβ2 and PLCβ3 [166, 218]. Gαq-GTP binding to PLC-β has been shown to cause a robust (upto three orders of magnitude) increase in the GTPase activity of Gαq [219]. Moreover, the crystal structure of PLCβ3 bound to activated Gαq reveals that a conserved module in the EF domain is involved in the enhancement of GTP hydrolysis [218]. The PH domain of PLCβ2 has also been shown to bind with high affinity to Rac GTPases [220, 221]. Although it has been reported that the EF domains of PLCs can bind Ca^{+2} there is little evidence for Ca^{+2}-mediated regulation of PLC activity via these domains [167].

PLCγ

PLCγ isozymes are activated by growth factor receptors like platelet-derived growth factor (PDGF), epidermal growth factor (EGF) or nerve growth factor (NGF). Two isoforms of PLCγ, PLCγ1 and PLCγ2, have been identified [166, 216]. They are ubiquitously expressed with highest levels of PLCγ1 detected in brain, lungs and thymus [222, 223] and of PLCγ2 in lungs, thymus and spleen [224, 225] (Table 3.4). PLCγ family is characterized by the presence of a highly structured region made up of two SH2 and one SH3 domain adjacent to a split PH domain that flanks the catalytic (X-Y) domain [226] (Fig. 3.9). Since SH2 and SH3 domains recruit effector proteins containing phosphorylated tyrosine residues and proline-rich (PXXP motif) sequences respectively, these domains enable PLCγ to form signaling complexes allowing for positive and negative modulation of their enzymatic activity, and thus to an isoform-specific signaling pathway [227, 228]. Moreover, studies show that the N-terminal PH domain of PLCγ can bind to phosphatidylinositol-4, 5-diphosphate, and to phosphatidylinositol-3,4,5-triphosphate respectively [212, 229, 230].

PLCδ

Three isoforms of PLCδ (PLCδ1, PLCδ3 and PLCδ4) have been identified. Of these PLCδ1 is expressed in brain, heart, lungs, testis, skeletal muscle and spleen; PLCδ3 is expressed in brain, skeletal muscle and heart; and PLCδ4 is expressed in sperm, testes, brain and skeletal muscle [166, 216] (Table 3.4). The structure of PLCδ is very similar to that of PLCβ although it lacks the PDZ binding domain (Fig. 3.9). In the case of the PH domain it has been reported that it is highly mobile in PLCδ1 and attached to the rest of the protein via a flexible link [212, 231]. Moreover, like in PLCγ, the N-terminal PH domain of PLCδ can bind to phosphatidylinositol-4,5-diphosphate and to phosphatidylinositol-3,4,5-triphosphate respectively [212, 229, 230]. Studies have indicated that PLCδ isoforms are sensitive to Ca^{+2} levels although not to the same extent as PLCζ and PLCη [166, 216]. A characteristic of PLCδ1 is that Ca^{+2} binding to the C2 domain promotes the translocation of the enzyme to the plasma membrane [232]. It is not clear whether other mammalian PLC isoforms are regulated by Ca^{+2} in a similar manner.

PLCε, PLCζ and PLCη

PLCε has been shown to be a direct effector of Ras protein. It is abundantly expressed in the heart, lungs and kidneys with relatively low or no expression in the brain [166, 233] (Table 3.4). PLCε family is characterized by the presence of isoform specific RasGEF and Ras binding domains through which the enzyme specifically interacts with the small GTPases, Ras, Rap or RhoA [163, 234] (Fig.3.9). Very little information is available about PLCζ, and η except that they are extremely sensitive to Ca^{+2} levels. PLCζ expression is restricted to sperm cells and this PLC isoform is characterized by the absence of the PH domain [235]. PLCη appears to be neuronal-specific and in the brain PLCη1 is abundantly expressed in hippocampus and Purkinje cell layer of the cerebellum while PLCη2 is abundantly expressed in hippocampus, cerebral cortex and olfactory bulb [166, 236, 237].

3.3.3 G PROTEIN-GATED INWARDLY RECTIFYING K^+ CHANNELS

G protein-gated inwardly rectifying K^+ (GIRK/Kir3) channels are activated by Gαi/o-coupled receptors [238]. Four GIRK subunits, GIRK1-4, come together to form homotetrameric or heterotetrameric GIRK channels [239]. Three channel subunits, GIRK1-3 exhibit broad distributions in the central nervous system, while GIRK4 expression is more restricted with highest levels being detected in the habenula [239, 240].

The major GIRK in the brain is composed of GIRK1 and GIRK2 subunits. Each GIRK heterotetramer subunit is made up of two transmembrane-spanning helices (TM1 and TM2) linked by an extracellular pore-forming region (H5) and cytoplasmic N- and C-terminal domains [241, 242] (Fig. 3.10). The H5 region serves as the "ion-selectivity filter" and has a sequence motif T-X-G-Y(F)-G found in other K^+-selective ion channels [241, 243]. GIRK channels lack an S4 voltage sensor region that is conserved in voltage-gated Na^+, Ca^{2+}, and K^+ channels thereby making them insensitive to membrane voltage [244]. Inward rectification, a defining characteristic of GIRK channels, is due to the block of outward K^+ flux by intracellular Mg^{2+} and polyamines [245, 246]. A high-resolution structure of the GIRK channel subunit shows that the channel pore measures ~30 Å in length and 7–15 Å in diameter and is composed of a wall of β-sheets surrounded by polar and charged residues. This creates a favorable environment for the binding of polyamines which blocks the pore giving rise to inward rectification [247]. The gating of GIRK channels is modulated by membrane PIP_2 levels and this interaction is regulated by sodium, intracellular pH, arachidonic acid and G proteins [248].

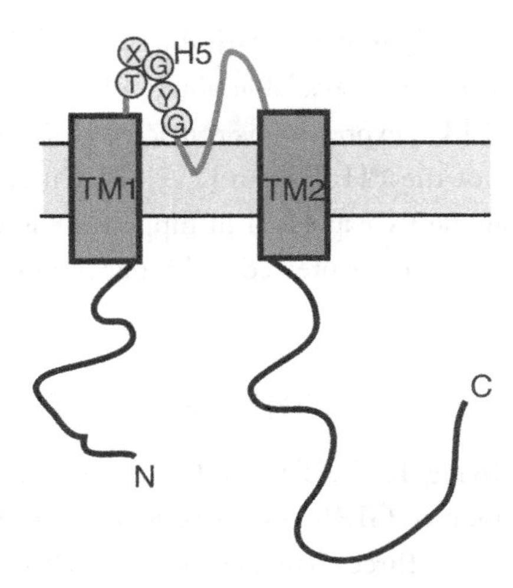

Figure 3.10. **Schematic representation of a single GIRK channel subunit.** Two transmembrane-spanning helices (TM1 and TM2) linked by an extracellular pore forming region (H5) form a single GIRK channel subunit. A complex of four such subunits form a functional GIRK channel.

Although studies have shown that GIRK channels are coupled to and activated by pertussis toxin-sensitive GPCRs, it is the Gβγ rather than Gα subunit that modulates channel activation [176]. This is supported by studies showing that sequestration of Gβγ inhibits GIRK channel activation by neurotransmitters [249]. Approximately 8-12 putative Gβγ-binding segments have been identified on the N- and C-terminal domains of the GIRK channel tetramer [250–254]. Cross-linking experiments suggest that one Gβγ binds to each GIRK channel subunit [255–257] at residues 34–86 on the N terminus and residues 318–374 and 390–462 on the C terminus [251, 258].

3.3.4 VOLTAGE-GATED CA^{2+} CHANNELS

Studies showed that activation Gαi/o-coupled receptors leads to the inhibition of voltage-gated Ca^{2+} channels [259-263]. Currently, these channels are classified based on their inhibition by specific toxins (Table 3.5). Thus, L-type channels characterized by high voltage activation, large single channel conductance and slow voltage-dependent inactivation are blocked by dihydropyridine [263–265]. T-type channels characterized by their transient opening, being activated at more negative membrane potentials followed by rapid inactivation can be blocked by kurtoxin [266–268]. N-type Ca^{+2} channels characterized by being more negative and faster than L-type but more positive and slower than T-type channels are blocked by ω-conotoxin GVIA. In addition, N-type Ca^{+2} channels are insensitive to L-type Ca^{+2} channel blockers [268]. R-type channels are characterized by blockade with SNX-482 while P/Q-type channels are blocked by ω-agatoxins [263] (Table 3.5).

Table 3.5. Characteristics of voltage-gated Ca^{+2} channels

Isoforms	Encoding gene for α subunit	Tissue distribution	Inhibitors	Physiological roles	References
L-type Ca+2 channels					
Cav1.1	*CACNA1S*	Skeletal muscle	dihydropyridine	Skeletal muscle contraction; transcriptional regulation	[263–265, 634, 635]
Cav1.2	*CACNA1C*	Neurons; cardiac & smooth muscle	dihydropyridine	Contraction of cardiac & smooth muscle; endocrine secretion; Ca^{+2} transients in neuronal cell bodies and dentrites; regulation of enzyme activity and of transcription	[634, 635]
Cav1.3	*CACNA1D*	Neuronal dendrites; sinoatrial node; cochlear hair cells	dihydropyridine	Endocrine secretion, cardiac pacemaking activity; Ca^{+2} transients in neuronal cell bodies and dendrites; auditory signal transduction	[634, 635]
Cav1.4	*CACNA1F*	Retina; sensory neurons		visual signal transduction	[634, 635]
P/Q-type Ca^{+2} channels					
Cav2.1	*CACNA1A*	Nervous system	ω-agatoxins	Neurotransmitter release; Ca^{+2} transients in neuronal dendrites	[263, 634, 635]
N-type Ca^{+2} channels					
Cav2.2	*CACNA1B*	Nervous system	ω-conotoxin CVIA	Neurotransmitter release; Ca^{+2} transients in neuronal dendrites	[268, 634, 635]

R-type Ca^{+2} channels					
Cav2.3	*CACNA1E*	Nervous system	SNX-482	Neurotransmitter release; Ca^{+2} transients in neuronal dendrites	[263, 634, 635]
T-type Ca^{+2} channels					
Cav3.1	*CACNA1G*	Neurons, cardiac muscle	kurtoxin	Pacemaking & repetitive firing	[266, 634, 635
Cav3.2	*CACNA1H*	Neurons, heart; adrenal gland	kurtoxin	Pacemaking & repetitive firing	[266, 634, 635]
Cav3.3	*CACNA1I*	Neurons	kurtoxin		[266, 634, 635

It is thought that Gα subunits confer specificity to receptor-mediated voltage-gated Ca^{2+} channel activation while the Gβγ subunits are responsible for direct modulation of channel activity [269–274]. This is supported by studies showing that expression of Gβγ mimics the effects of neurotransmitters and guanyl nucleotides on N-type and P/Q-type Ca^{2+} channels while expression of Gα subunits does not [270, 275]. In addition, Gβγ has been shown to directly interact with and inhibit N-type Ca^{2+} channels and stimulate L-type Ca^{2+} channels [269–273]. Sites of interaction between the Gβγ subunit and voltage-gated Ca^{2+} channels include the N-terminal, C-terminal, and α-interaction domain in the α1 channel subunit [276] (Fig. 3.11), and the guanylyl kinase binding domain in the β channel subunit [276].

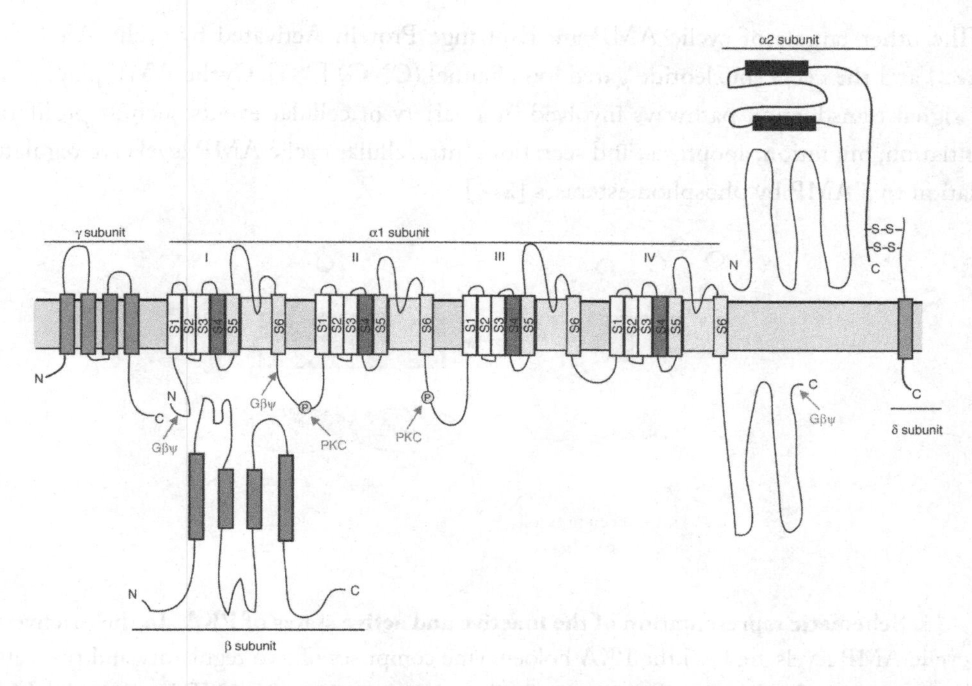

Figure 3.11. **Schematic organization of voltage-gated Ca^{+2} channels.** These channels comprise several subunits: α1, α2δ, β, and γ. The voltage sensor domain is formed by the charged S4 segment (purple) and the pore region by the S5 to S6 segments (yellow). Sites of interaction with Gβγ and of phosphorylation by PKC shown.

3.4 DOWNSTREAM EFFECTORS OF G PROTEIN-MEDIATED SIGNALING

3.4.1 CYCLIC AMP

One of the well-established intracellular second messengers is cyclic AMP. Cyclic AMP was initially identified in 1957 by Sutherland and co-workers while studying the regulation of glycogen metabolism by adrenaline [277]. Activation of Gαs-coupled receptors induces an increase while of Gαi-coupled receptors a decrease in intracellular cyclic AMP levels. One of the main targets of cyclic AMP are protein kinases called cAMP-dependent protein kinases or PKA [278]. These kinases are composed of two regulatory and two catalytic subunits; the regulatory subunits bind to cAMP and this leads to the release of the catalytic subunits that become active and phosphorylate substrate proteins [279, 280] (Fig. 3.12).

The other targets of cyclic AMP are Exchange Protein Activated by cyclic AMP (Epac) [281, 282] and the cyclic-nucleotide gated ion channel (CNG) [283]. Cyclic AMP plays a crucial role in signal transduction pathways involved in a variety of cellular events such as proliferation, differentiation, migration, apoptosis, and secretion. Intracellular cyclic AMP levels are regulated by degradation to 5'AMP by phosphodiesterases [284].

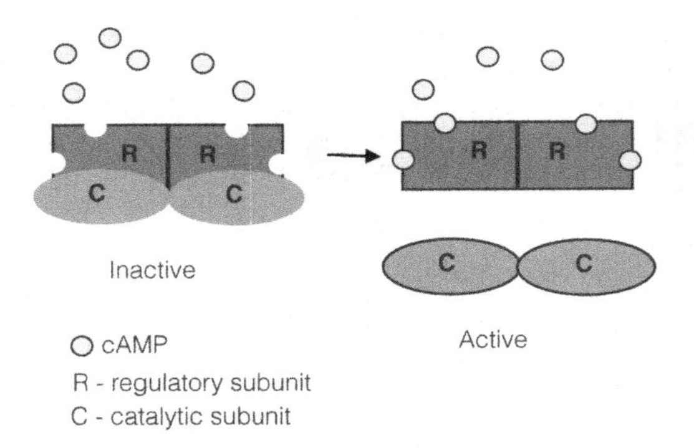

Figure 3.12. **Schematic representation of the inactive and active states of PKA.** In the inactive state (when cyclic AMP levels are low) the PKA holoenzyme comprises of two regulatory and two catalytic subunits. Activation of a Gαs-coupled receptor leads to increases in cyclic AMP levels; cyclic AMP binds to the regulatory subunits of PKA causing them to dissociate from the catalytic subunits; the latter thus became active and can phosphorylate downstream PKA substrates.

3.4.2 CALCIUM AND PHOSPHOINOSITIDES

Other well-studied intracellular second messenger systems are those released by the hydrolysis of phospholipids by PLC. Activation of Gαq/11-coupled receptors leads to the activation of PLC, which in turn hydrolyzes the membrane phospholipid, phosphatidylinositol-4,5-bisphosphate also known as PIP_2 into diacylglycerol (DAG) and inositol-1,4,5-triphosphate (IP3). DAG, retained in the membrane, activates PKC. IP3 is a soluble molecule that diffuses to bind to a ligand gated calcium ion channel called the IP3 receptor. This, in turn, leads to the opening of the channel releasing calcium to the cytoplasm. The released calcium binds to a protein called calmodulin that, in turn, activates a kinase called calcium-calmodulin-dependent kinase (CAMK) [161, 166, 167]. As with the cyclic AMP system, the levels of phosphoinositides are regulated by the enzymes that modify them and the level of calcium is regulated by calcium pumps that actively remove calcium from the cytoplasm [161, 166, 167]. Due to the activation of multiple kinases, the end result of activation of the phosphoinositide signaling pathway is much more complex than the cyclic AMP pathway since different cell types may contain a different repertoire of kinases that could lead to the phosphorylation of a distinct set of substrates. For example, Gαq/11-coupled receptors can activate the Ca^{2+}-

and PKC-regulated kinase, Pyk2, and this promotes the Ras-dependent activation of the MAPK cascade [285]. In addition, the Ras family GTPase, Ral, may be activated by Gαq/11-coupled receptors via a pathway involving the binding of Ca^{2+}-calmodulin to the C-terminus of Ral A [286].

3.5 GPCR DESENSITIZATION AND DOWN-REGULATION

Following continued exposure to receptor agonists, GPCR-mediated signaling is desensitized. A common mechanism of desensitization of neuropeptide receptors is by the phosphorylation of the receptor by GRKs. GRKs can discriminate between the inactive and agonist-activated states of the receptor, in part, because they are catalytically activated by stimulated receptors [287]. Receptor phosphorylation leads to the recruitment of β-arrestins to the receptor; this is then followed by clathrin-coated pit-mediated endocytosis of the receptor (Fig. 3.13). Endocytosed receptors are either recycled back to the cell surface to undergo another round of signaling or are targeted to lysosomes for degradation (Fig. 3.13). In the latter case this leads to receptor down-regulation by decreasing the number of receptors available to bind agonists. In the following sections we will describe these processes with a focus on G-protein receptor kinases and β-arrestins.

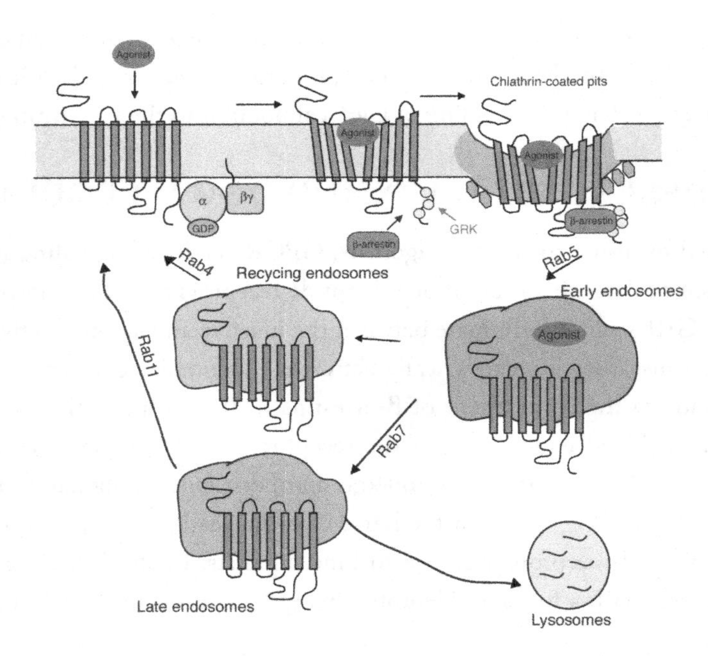

Figure 3.13. **GPCR endocytosis and recycling.** Following activation, GPCRs are phosphorylated by GRKs and this leads to recruitment of β-arrestin to the phosphorylated receptor. The β-arrestin bound GPCR is targeted to clathrin-coated pits for endocytosis and desensitization of G protein-mediated signaling. Endocytosed receptors in early endosomes can undergo a rapid or slow recycling to the cell surface so as to undergo another round of signaling events. Alternatively endocytosed receptors in early endosomes are trafficked to late endosomes and ultimately lysosomes where they are degraded leading to down-regulation of cell surface receptors. GRK, G protein-coupled receptor kinase; Rab4, rapid recycling compartment marker; Rab5, early endosome marker; Rab7, late endosome marker; Rab11, slow recycling compartment marker.

3.5.1 G PROTEIN-COUPLED RECEPTOR KINASES

G protein-coupled receptor kinases (GRKs) are serine/threonine kinases involved in GPCR signal desensitization. To date seven GRKs have been identified in humans that are encoded by different genes (Table 3.6). GRKs are divided into three subfamilies based on sequence homology and tissue expression: rhodopsin kinase (GRK1 and GRK7), β-adrenergic receptor kinase (GRKs 2 and 3) and GRK4 kinase (GRK4, GRK5, and GRK6) [288-290]. The expression of rhodopsin kinases and GRK4 is restricted to retina and testes, respectively, while the other GRKs are ubiquitously expressed (Table 3.6) [288–290]. An interaction between a GRK and the activated GPCR on the plasma membrane is required for GRK-catalyzed receptor phosphorylation.

Table 3.6. Characteristics of non-visual GRK isoforms

Isoforms	Encoding gene	Tissue Distribution	Regulator	References
GRK2	*ADRBK1*	Brain; heart; hematopoietic cells; skeletal muscle	α-actinin; Ca^{+2}-calmodulin; caveolin; Gαq/11; Gβγ; PIP_2; phosphatidylserine; PKC; ERK 1/2; c-Src	[194, 293, 294, 296, 636–649]
GRK3	*ADRBK2*	Brain; skeletal muscle; olfactory epithelium	α-actinin; Ca^{+2}-calmodulin; caveolin; Gαq/11; Gβγ; PIP_2	[293, 294, 296, 636, 639–644, 650]
GRK4	*GRK4*	Brain; lungs; kidneys; testis	α-actinin; Ca^{+2}-calmodulin; PIP_2	[293, 636, 639–641, 644, 651]
GRK5	*GRK5*	Brain; heart; hematopoietic cells; kidneys	α-actinin; Ca^{+2}-calmodulin; caveolin; PIP_2; PKC	[636-641, 644-646, 651]
GRK6	*GRK6*	Brain; hematopoietic cells; lungs; skeletal muscle	α-actinin; Ca^{+2}-calmodulin; caveolin; PIP_2	[293, 636-641, 644]

ERK1/2, extracellular signal-regulated kinases 1/2; PIP_2, phosphatidylinositol 4,5-bisphosphate; PKC, protein kinase C

GRKs consist of a single polypeptide chain of 60-80 kDa that is organized into a central catalytic domain, flanked by a N-terminal region containing a RGS homology domain and a variable length C-terminal region [291] (Fig. 3.14). The N-terminal region appears to play a role in receptor recognition and anchoring to intracellular membranes [292, 293] while the C-terminal domain dictates subcellular localization and membrane association/translocation of GRKs [294, 295]. GRKs involved in phosphorylating neuropeptide receptors are GRK2 and GRK3 also known as β-adrenergic receptor kinases [291]. GRK2 and GRK3 translocate to the plasma membrane upon agonist stimulation and both kinases have a pleckstrin homology domain through which they interact with PIP_2 or other acidic phospholipids at the plasma membrane and this affects kinase activity [291, 296, 297]. For example, phosphatidylserine and PIP_2 increase by 2–3 fold the phosphorylation of membrane receptors by GRK2 [297]. The membrane translocation of GRK2 and GRK3 requires the activation of G proteins since GRK2 and GRK3 must bind free Gβγ subunits anchored to the membrane in order to be recruited to the membrane [291]. Furthermore, the interaction of GRK2 and GRK3 with the Gβγ stimulates the phosphorylation of GPCRs [291].

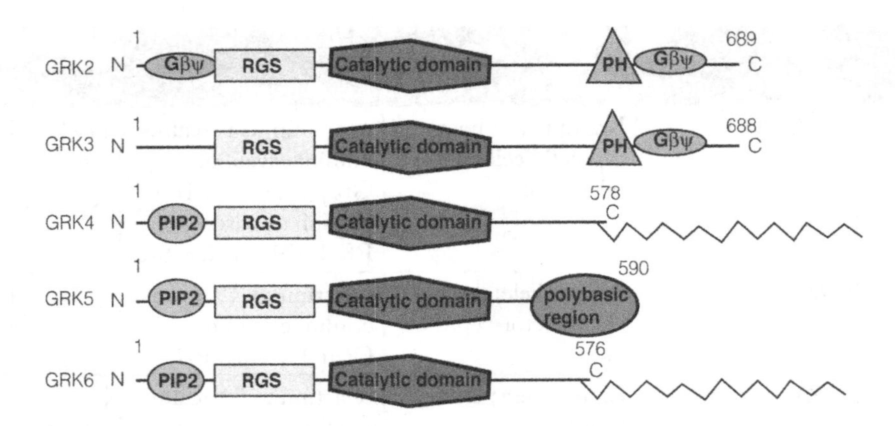

Figure 3.14. **Structure of non-visual GRKs.** Non-visual GRK isoforms are characterized by the presence of a catalytic domain and a RGS homology domain. The different isoforms differ in the length of the C-terminus, in post-translational lipid modifications and presence of PH, Gβγ, and PIP_2 binding domains. RGS, regulator of G-protein signaling; PH, pleckstrin homology domain; PIP_2, phosphatidylinositol- 4,5-bisphosphate.

Studies show that the activity of GRK2 can be modulated by phosphorylation. For example, ERK1/2 can phosphorylate GRK2 and this causes a decrease in the kinase activity of GRK2 as well as in its binding to Gβγ [298]. In addition, GRK2 can also be phosphorylated at tyrosine residues in the RGS domain following receptor activation. This phosphorylation depends on the ability of β-arrestin to bind to and recruit c-Src to the receptor. c-Src phosphorylates GRK2 which increases the activity of GRK2 toward receptor substrates and potentiates receptor desensitization. Other kinases that can modulate the activity of GRKs include PKA and PKC [291]. PKA can phosphorylate GRK2 increasing its affinity for Gβγ subunits as well as promoting its translocation to the membrane [291]. PKC can phosphorylate GRK2 and GRK5 [291]; phosphorylation of GRK2 stimulates GRK-mediated receptor phosphorylation, probably through an increase in the translocation of GRK2 to the membrane, whereas phosphorylation of GRK5 by PKC inhibits GRK-mediated receptor phosphorylation as well as the translocation of GRK5 to the membrane [291].

3.5.2 β-ARRESTINS

Phosphorylation of GPCRs by GRKs leads to the recruitment and binding of β-arrestin to the phosphorylated C-terminal tail (Fig. 3.13). Arrestins recognize both GRK phosphorylation sites on the receptor and the active conformation of the receptor, which together drive robust arrestin association [73, 299].

To date four functional arrestin gene family members have been identified which are encoded by the *SAG*, *ARRB1*, *ARRB2*, and *ARR3* genes [299]. Two of these are expressed only in the retina (visual arrestin and cone arrestin) and regulate photoreceptor function [299]. The non-visual arrestins, β-arrestin 1 and β-arrestin 2 are expressed ubiquitously in all cells and tissues and function in the desensitization of most neuropeptide receptors [299]. β-arrestin 1 and 2 exhibit 78% homology and contain binding motifs for clathrin and the β2-adaptin subunit of the AP-2 complex in their C-terminal tail which allows them to function as adaptor proteins and target GPCRs to clathrin-coated pits for endocytosis [299].

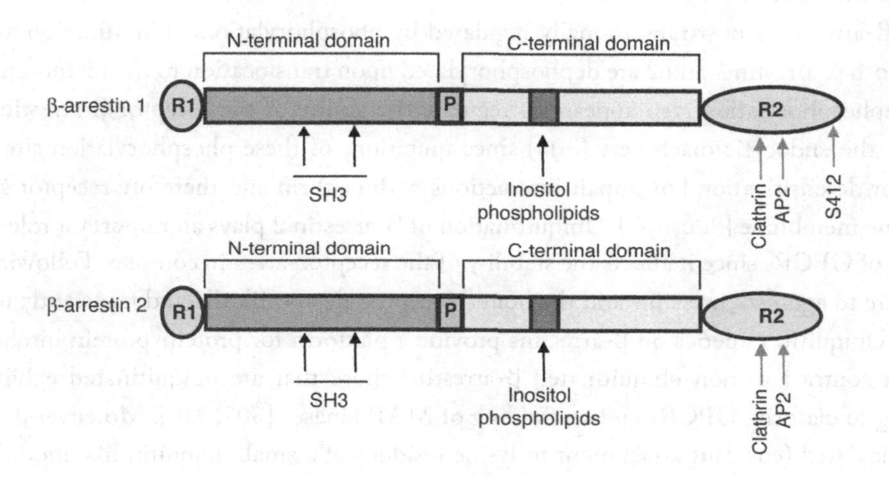

Figure 3.15. **Structure of non-visual arrestins.** β-arrestins are characterized by the presence of a seven stranded β sandwich N-terminal and C-terminal domains separated by a phosphate sensor domain (P) in addition to a N-terminal regulatory domain (R1). The length of the C-terminal regulatory domain (R2) differs between the β-arrestin 1 & 2 isoforms. Red arrows indicate (i) primary site of β-arrestin 1 phosphorylation (S412), (ii) binding motif for clathrin (LIEF) and (iii) binding motif for AP2 (RXR). Black arrows indicate (i) c-Src SH3 binding domain (PXXP) and (ii) the recognition domain for inositol phospholipids.

The structure of arrestins comprise two major domains, an N-terminal and a C-terminal domain each of which is composed of a seven stranded β sandwich [300–302]. These two domains are connected by a polar core, termed the phosphate sensor domain, that in the basal state is embedded between the N- and C-terminal domains [300–302] (Fig. 3.15). Residues from the free N- and C-terminal tails contribute to the formation of this polar core in unbound arrestin keeping it in an inactive state [300–302]. The polar phosphate sensor domain plays an important role in the conformational changes that occur following the binding of arrestin to a phosphorylated receptor [300–302]. It has been proposed that when arrestin binds to a phosphorylated receptor, the receptor tail penetrates the polar core and displaces the C-terminal tail of arrestin from it; this allows conformational changes that permit tight binding of arrestin to the receptor [303]. Support for this comes from studies showing that mutations that disrupt intramolecular interactions in the polar core result in arrestins that bind with high affinity to agonist occupied GPCRs without the requirement of receptor phosphorylation [304, 305]. The N-terminal domain contains sites that can distinguish between phosphorylated and non-phosphorylated GPCRs [303]. The structure of β-arrestins also contains a cationic amphipatic helix that serves as a reversible membrane anchor for interaction with GPCRs; this would facilitate the formation of high affinity complexes between arrestin and activated GPCRs [303].

β-arrestin is posttranslationally regulated by phosphorylation, ubiquitination and sumoylation. Both β-arrestin 1 and 2 are dephosphorylated upon translocation to the membrane [306, 307]. The dephosphorylation step appears to regulate the ability of the receptor-β arrestin complex to engage the endocytic machinery [306] since mutations of these phosphorylation sites can induce receptor desensitization but impair interactions with clathrin and therefore receptor sequestration from the membrane [306, 307]. Ubiquitination of β-arrestin 2 plays an important role in the endocytosis of GPCRs since it affects the stability of the receptor-arrestin complex. Following continued exposure to agonists, arrestins and the bound receptor are specifically and transiently ubiquitinated [308]. Ubiquitin moieties on β-arrestins provide a platform for protein-protein interactions given that in contrast to non-ubiquitinated β-arrestins those that are ubiquitinated exhibit enhanced binding to clathrin, GPCRs and scaffolding of MAP kinases [309, 310]. Moreover β-arrestins can be sumoylated (covalent attachment to lysine residues of a small ubiquitin like modifier, SUMO). This posttranslational modification plays an important role in interaction with AP2 and receptor endocytosis via clathrin-coated vesicles but does not affect the affinity of β-arrestin interaction with GPCRs [311].

β-arrestin and GPCR endocytosis

Arrestin bound receptors exist as relatively stable complexes. In this state, β-arrestins can act as adapter proteins that physically link the receptor to the clathrin mediated endocytic machinery thereby promoting receptor endocytosis via clathrin-coated pits (Fig. 3.13). Support for the involvement of β-arrestins in GPCR endocytosis comes from studies using siRNA to β-arrestins and β-arrestin 1 and 2 double knockout animals [312, 313]. Moreover, overexpression of β-arrestin was shown to increase GPCR internalization and to rescue the internalization of mutant non-internalizing receptors [314].

GPCRs have been classified as class A or B receptors based on the isoform of arrestin associated with the receptor and on the stability of this interaction [310, 315]. Class A receptors like the β2-adrenergic or the μ opioid receptor are characterized by (i) higher binding affinity to β-arrestin 2 than to β-arrestin 1 and do not bind to visual arrestin; (ii) ubiquitination of receptor associated arrestin is transient and leads to the formation of transient signaling complexes at the cell surface;

(iii) rapid recycling and resensitization dynamics [310, 315, 316]. Class B receptors like the angiotensin II1a or the vasopressin 2 receptor are characterized by (i) binding with similar affinities to β-arrestin 1 and 2 as well as to visual arrestin; (ii) sustained ubiquitination of β-arrestin leading to the formation of tight GPCR-arrestin complexes that co-internalize in endocytic vesicles; (iii) sustained MAPK activity and slow recycling and resensitization dynamics [316].

3.6 G PROTEIN-INDEPENDENT SIGNALING

G protein-independent signaling occurs following recruitment of β-arrestin to the phosphorylated receptor. Studies showed that receptor associated β-arrestins could serve as scaffolds for the recruitment of new signaling molecules to the receptor leading to the activation of a variety of signaling cascades [73] such as MAPKs, the small GTPase RhoA, Ral GDP dissociation factor and cofilin, as well as in the inhibition of signaling molecules such as NF-kB and LIMK (for review see [317]).

The best characterized role of β-arrestin-dependent signaling is in the regulation of MAPKs. Studies show that ERK1/2 (p42/44MAPK), p38MAPK and JNK can be activated by β-arrestin-dependent signals [318]; this leads to either the cytosolic sequestration [319, 320] or to the nuclear translocation [321] of the activated kinase. Although the ability of β-arrestin to induce GPCR endocytosis may play a role on its ability to activate MAPKs [321, 322], for some receptors β-arrestin-dependent ERK1/2 activation occurs in the absence of receptor internalization [323, 324]. The latter observation is supported by studies that used an inducible strategy to target β-arrestin to the vasopressin receptor without receptor activation and showed that this led to membrane localization of Raf, MEK1/2 and ERK1/2 via the β-arrestin scaffold and was sufficient to trigger ERK1/2 activation [325]. For some receptors β-arrestins scaffold tyrosine kinases such as Src, thus linking signaling to classical Ras-dependent pathways of ERK1/2 activation [317]. In addition, studies with a GPCR/β-arrestin chimera showed that β-arrestin-dependent ERK1/2 activation required PKA and PKC activity [317]. This suggests that β-arrestins scaffold and activate a number of enzymes linked to MAPK activity, independent of agonist-induced receptor activation. These proteins can be distinct from those activated by G protein-dependent pathways; alternatively, β-arrestins can effectively highjack proteins from the G protein pathway to direct alternate signaling responses [317].

3.7 β-ARRESTIN-DEPENDENT SIGNALING AND BIASED AGONISM

The observation that different ligands could differentially activate the repertoire of signaling cascades suggested that GPCRs could exist in multiple active conformations and that ligand binding could promote/stabilize one of these conformations, a phenomenon described as ligand-directed trafficking of receptor signaling also known as functional selectivity or biased agonism [326–329]. This would suggest the possibility that different active receptor conformations are responsible for the alternative coupling to G proteins or β-arrestins.

Studies examining biased agonism by β-adrenergic receptor ligands found that some compounds displayed similar efficacies toward both G protein-mediated and β-arrestin-mediated signaling pathways, while some compounds that were inverse agonists in the G protein-mediated assay (adenylyl cyclase activity) acted either as agonists or neutral antagonists of the β-arrestin pathway (ERK1/2 phosphorylation) and vice-versa [326]. Studies with the parathyroid hormone receptor also showed that administration of the receptor agonist PTH(1-34), which activates G protein-dependent signaling, and of PTH-βarr, that activates a β-arrestin dependent signaling, stimulated anabolic bone formation in mice. However, PTH-βarr does not induce hypercalcemia or an increase in markers for bone resorption [329]. This suggests that biased agonism could be of therapeutic benefit to treat diseases involving GPCRs. Thus biased receptor agonists could improve pharmacological specificity and efficacy by selectively activating the pathways that are relevant to the treatment of a given disorder.

In recent years it has been shown that receptor heteromerization can also contribute to biased signaling. A study examining signaling by μ-δ heteromers reported that while homomers of the μ opioid receptor mediate signaling via G proteins, μ-δ heteromers signal via β-arrestin [330]. Activation of μ homomers leads to a G protein-mediated increase in the phosphorylation of ERK1/2 that is rapid and transient; phosphorylated ERK1/2 translocates to the nucleus where it phosphorylates nuclear ERK1/2 substrates [330]. In contrast, activation of μ-δ heteromers leads to a sustained increase in the phosphorylation of ERK1/2 via a β-arrestin-dependent pathway, the phosphorylated ERK1/2 does not translocate to the nucleus and phosphorylates cytoplasmic ERK1/2 substrates [330]. Thus the status of the receptor (homomer versus heteromer) appears to stabilize its conformation so as to signal either via a G protein- or a β-arrestin-mediated pathway. Thus heteromerization can also contribute to biased agonism by modifying the kinetics and spatio-temporal dynamics of signaling.

3.8 MODULATION OF GPCR SIGNALING BY HETERODIMERIZATION

Protein-protein interactions in the form of "receptor heteromerization" can modulate the signaling and functional properties of a receptor [331–333] (Fig. 3.16). Over the last decade studies have shown that an increasing number of GPCRs including those for neuropeptides form heteromers that exhibit distinct properties from receptor homomers (for reviews see [331, 333–338]). These include opioid, cannabinoid and dopamine receptors for which an increasing number of heteromers have been described (Table 3.7). A few examples of heteromers involving neuropeptide GPCRs are described below.

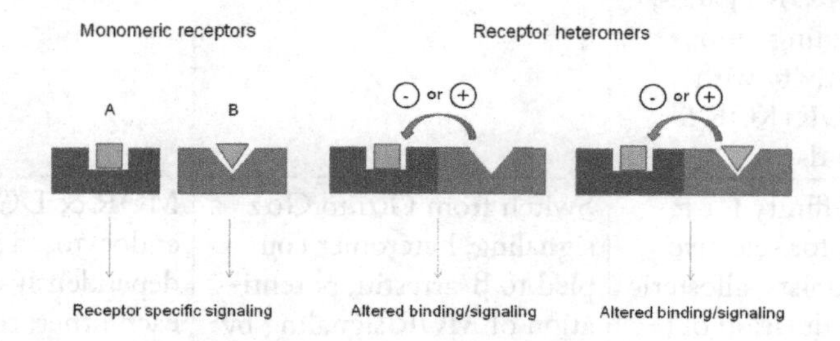

Figure 3.16. Modulation of GPCR signaling by heteromerization. (Left panel) Activation of receptor A or receptor B (which exist as monomers, shown, or as homodimers) by receptor-specific ligands can lead to receptor specific signaling. However, association of receptor B with receptor A, through heterodimerization, can lead to alterations (inhibition or enhancement) in the binding of ligands to receptor A and signaling through receptor A. Additionally, binding of receptor-specific ligands to receptor B may alter receptor A binding and signaling.

Table 3.7. Heteromers involving opioid, cannabinoid and dopamine receptors.

Heteromer pair	Changes in heteromer properties			References
	Binding	Signaling	Trafficking	
DOR-α_{2A}AR	Not reported	Synergy in inhibition of neuropeptide release; ↑DOR-mediated neurite outgrowth	Not reported	[652, 653]
DOR-β_{2A}R	Not changed	Not changed	Internalization of β_2AR by DOR agonists	[654, 655]
DOR-CB_1R	Not reported	↓ CB_1R-mediated signaling	Change in CB_1R localization	[656, 657]
DOR-CXCR4	Not changed	Non-functional heteromer	Not changed	[658]
DOR-D1R	Not reported	D1R agonists↓ DOR signaling	DOR redistribution by cocaine withdrawal	[659, 660]

DOR-KOR	↓ affinity for selective DOR or KOR ligands; binding co-operativity with DOR+KOR ligands	DOR + KOR agonists↑ signaling	Altered DOR trafficking	[654, 661]
DOR-MOR	↓ affinity for receptor selective agonists; allosteric modulation of MOR binding by DOR ligands and *vice-versa*	Switch from Gαi to Gαz signaling; heteromer coupled to β-arrestin; potentiation of MOR signaling by DOR ligands and *vice-versa*	MOR & DOR endocytosed independently of each other; some DOR agonists induce heteromer internalization	[50, 330, 339–342, 597, 662–666]
DOR-SNSR	Not reported	Preferential Gαq signaling	Not reported	[343]
KOR-β_2AR	Not changed	↓ β_2AR signaling	Altered	[655]
KOR-APJ	Not reported	↑ signaling by agonists to either receptor	Not reported	[667]
MOR-α_{2A}AR	Not reported	α_{2A}R↑ MOR and α_{2A}R ligands ↓ MOR signaling	Agonist to one receptor does not cause heteromer internalization	[668–670]
MOR-CB_1R	Not reported	CB_1R ↓ MOR and MOR ↓CB_1R signaling	Not reported	[657, 671]
MOR-CCR5	Not reported	MOR agonist ↓CCR5 and CCR5 agonist ↓MOR signaling	Not reported	[672, 673]
MOR-GRPR	Not reported	MOR-mediated Ca^{+2} signaling seen only in cells expressing the heteromer	MOR-mediated GRPR internalization	[674]
MOR-KOR	↓ affinity for MOR selective ligands	Not reported	Not reported	[52, 661, 675]
MOR-NK1	Not changed	Not changed	MOR internalization by NK1 agonist & *vice-versa*	[344]
MOR-OFQ/N	↑affinity for MOR selective ligands	↓ potency of MOR ligands	Not reported	[52, 676]

MOR-SSTR2	Not changed	Not changed	↑ heteromer internalization by SSTR2 agonist	[677]
CB_1R-$A_{2A}R$	Not reported	CB1R antagonist ↓$A_{2A}R$ and $A_{2A}R$ antagonist ↓ CB_1R signaling	not reported	[678]
CB_1R -$\beta_{2A}R$	Not reported	Alteration in signaling of individual receptors	CB_1R internalization by β2AR agonist	[679]
CB_1R-AT1R	Not reported	AT1R signals via Gαi instead of Gαq; CB_1R agonists ↑ AT1R and CB_1R antagonists ↓ AT1R signaling	Not reported	[599]
CB_1R-CB_2R	Not reported	Antagonistic interactions	Not reported	[680]
CB_1R-D2R	Not reported	CB_1R signals via Gαs in presence of D2R agonist	Not reported	[681-684]
CB_1R –OX1R	Not reported	CB_1R ↑OX1R signaling	OX1R agonist ↑ heteromer internalization	[685, 686]
D1R-D2R	Not changed	Gαq-mediated signaling by the heteromer	Not reported	[687–692]
D1R-D3R	↑ dopamine binding	↑ dopamine signaling	Heteromer internalized only by agonists to both receptors	[693]
D1R-$A_{1A}R$	Not reported	↓ D1R signaling by combination of agonists to both receptors	Not reported	[694]
D2R-D3R	Not reported	↑ potency of preferential protomer agonists	Not reported	[695, 696]
D2R-$A_{2A}R$	Not reported	D2R↓ efficacy & potency of $A_{2A}R$ agonists	Heteromer internalization by agonist to either protomer	[697–701]
D2R-GSHR1a	Not reported	D2R signals via Gαq and not Gαi	Not reported	[702]
D2R-NTS1	↓ affinity for D2R agonist in presence of NTS1 agonist	↑potency of D2R in presence of NTS1 agonist	NTS1 agonist ↑ D2R internalization	[703, 704]

D2R-OTR	↑ D2R antagonist binding by low doses of OTR agonist	↑efficacy and potency of D2R signaling by OTR agonist	Not reported	[705]
D2R-SSTR2	Modulation of binding properties of individual protomers	SSTR2 agonist ↑D2R signaling	D2R agonist↑ SSTR2 internalization	[706]
D2R-SSTR5	↑ affinity for each protomer agonist	Agonist to one protomer ↑ signaling by agonist to second protomer	Not reported	[707]
D2R-5-$HT_{2A}R$	D2R agonist ↑affinity of 5-$HT_{2A}R$ agonist	D2R agonist ↓potency of hallucinogenic 5-$HT_{2A}R$ agonist	Not reported	[708]
D4R-$\alpha_{1B}AR$	D4R agonist ↓binding displacement by $\alpha_{1B}AR$ agonist	D4R agonist ↓$\alpha_{1B}AR$ signaling	Not reported	[709]
D4R-β_1AR	D4R agonist ↓ binding displacement by β_1AR agonist	D4R agonist ↓ β_1AR signaling	Not reported	[709]

As described in section 3.7 heteromerization between μ and δ opioid receptors modulates the signaling properties of individual receptors [50, 339–341]. While μ or δ opioid receptors signal through the classical Gαi/o-mediated pathway, the μ-δ heteromers signals either via a Gαz- or a β-arrestin-mediated signaling pathway [330, 342]. Interestingly, μ-δ heteromers were found to be up-regulated in the brain and spinal cord following chronic morphine administration. In another study heteromerization between δ opioid receptors and CB1 cannabinoid receptors was found to be enhanced during neuropathic pain. In this case while CB1 cannabinoid receptors and δ opioid receptors transduce signaling via Gαi/o-mediated pathways CB1 cannabinoid-δ opioid heteromers transduce signals via a β-arrestin-mediated pathway. This suggests that heterodimerization could serve as an additional mode of regulation of receptor activity that could be differentially altered during pathology.

An interesting case is that of heteromers between δ opioid receptors (Gαi/o-coupled) and sensory neuron-specific receptors (Gαq-coupled). An endogenous peptide, BAM-22, can bind to sensory neuron-specific receptors via its C-terminal region and elicit Gαq-mediated signaling leading to the activation of phospholipase C. The same peptide binds to δ opioid receptors via its N-terminal region to elicit Gαi/o-mediated signaling leading to the inhibition of adenylyl cyclase activity. However, heteromerization between δ opioid receptors and sensory neuron-specific receptors promotes a BAM22-mediated SNSR-4 signaling (inositol phosphate production) that is sensitive to the δ opioid receptor selective antagonist naltrexone [343].

Studies have also described heteromerization between substance P receptors and μ opioid receptors. In this case the selective μ opioid receptor agonist, DAMGO, promotes the phosphorylation and internalization of substance P receptors (NK1 receptors) [344]. Conversely, treatment with substance P promotes the phosphorylation and internalization of μ opioid receptors [344]. Studies show that heteromerization between neuropeptide Y1 (NPY1) and NPY5 receptors leads to altered agonist and antagonist responses and in a reduction in receptor internalization. Selective NPY5 receptor agonists produce a greater inhibition of adenylyl cyclase activity in cells expressing the heteromers compared to cells expressing only NPY5 receptor homomers. In addition, in cells expressing NPY1-NPY5 heteromers the NPY1 receptor antagonist does not block signaling by a non-selective receptor agonist, while the NPY5 receptor antagonist exhibited uncompetitive kinetics [345]. Studies examining heteromerization between vasopressin V1b receptors and corticotropin releasing hormone receptor 1 (CRHR1) show that vasopressin can potentiate CRH-mediated increases in cyclic AMP levels and that CRH can potentiate vasopressin-mediated increase in inositol phosphate levels only in cells co-expressing both receptors [346]. Taken together these studies indicate that receptor heteromerization increases the signaling repertoire of the ligands to individual receptors.

CHAPTER 4

Neuropeptide Processing and Regulation

4.1 INTRODUCTION

Neuropeptides are biologically active peptides resulting from the processing of large precursor proteins in the brain and neuroendocrine tissues. These peptides range in length from ~3 to 40 amino acid residues. As a rule a peptide is considered to be a neuropeptide if (i) it is synthesized in the brain; (ii) it is secreted from brain cells at physiological levels; (iii) either its synthesis or secretion is regulated; and (iv) it influences the function of another cell [347].

Functional neuropeptides are the result of the proteolytic processing of large precursor proteins. The specificity of this processing is essential to the function of a neuropeptide since even a minor defect in processing can result in peptides having diminished activity, or worse, that are non-functional [348]. Moreover, different processing enzymes can produce functionally distinct peptides from a single precursor. A case in point are the opioid peptides. This diverse set of neuropeptides arises from the processing of one of three precursors: prodynorphin, proenkephalin, or POMC [349]. These proproteins are differentially processed by peptidases of the regulated neuroendocrine secretory pathway, and give rise to a large number of biologically active peptides with different activities [350, 351]. β-endorphin is an especially interesting example of the control of peptide function by differential cleavage: the longer form of the peptide is an opioid receptor agonist but the shorter form, lacking its last four residues, is an antagonist at the same receptor [352]. Thus, neuropeptide processing plays an integral role in generating the bioactive forms of the peptides, while the regulation of peptide processing dictates the physiological function of the product. In the following sections we will describe how neuropeptides are processed from large precursors, sorted into secretory vesicles, secreted and regulated.

4.2 NEUROPEPTIDE PROCESSING

Neuropeptides are synthesized as proproteins in the rough endoplasmic reticulum by membrane bound ribosomes. The precursors are initially synthesized with an N-terminal signal peptide (preproprotein) that helps to target them to the lumen of the endoplasmic reticulum. Thus it is the cleavage of this N-terminal signal peptide that is the first of multiple cleavages steps that will ultimately yield a mature peptide. After synthesis and cleavage of the preproprotein, the proprotein transits from the cis- to the *trans*-Golgi network, during which required glycosylation steps occur, in much the same manner as for other proteins [353]. Peptidergic vesicles are formed when the proprotein exits the *trans*-Golgi network (TGN). It is at this point where the earliest endopeptidic cleavages occur. In order to generate mature, active signaling neuropeptides, the proproteins undergo specific, limited proteolytic processing by endopeptidases and carboxypeptidases, as well as other post-translational modifications. The proteolytic cleavage of neuropeptide precursors occurs in a tissue specific or cell-type specific fashion, where the precursors may be cleaved at different residues depending upon the tissues in which they are found, and the native processing enzymes present in these tissues [354].

In the classical neuropeptide processing scheme, endopeptidases cleave the proprotein on the carboxy-terminal side of either paired basic residues, or tetrabasic residues with consensus motifs of [R/K]– [X]n– [R/K] where X represents any amino acid residue, R/K the basic amino acid residues arginine and lysine, and n (the number of spacer amino acid residues) is 0, 2, 4, or 6 [355–357]. The initial proteolytic step is carried out at basic sites by subtilisin/kexin-like prohormone convertases (PC), a family of enzymes that structurally resembles the subtilisins of bacteria, or the kexins of yeast [358]. The PC family includes furin, PC1/3, PC2, PC4, PACE4, PC5/6, and PC7 [359] (Table 4.1). After proteolysis at basic sites, the carboxyl-terminal basic amino acids of resulting peptide intermediates are trimmed off by specialized metallocarboxypeptidases (carboxypeptidases E and D) leading to the formation of mature peptides [360].

Table 4.1. Prohormone Convertases

Enzyme	pH Optimum	Localization	References
PC1/3	5–6	Dense Core Vesicles	[710]
PC2	5–6	Dense Core Vesicles	[710]
Furin	6–7	*Trans*-Golgi Network	[711]
PC4	7	Cell Surface	[712]
PC5/6	7	Cell Surface	[712]
PACE4	7	Cell Surface	[713]
PC7	6-7	*Trans*-Golgi Network	[713]

The peptidases of the regulated secretory pathway have different pH optima and Ca^{2+} dependent activities, which help determine where they are active [361]. As peptides transit the regulated secretory pathway, from the TGN to immature secretory vesicles, and then mature large dense-core vesicles (LDCVs), there is a corresponding decrease in the intravesicular pH, from neutral in the TGN (pH ~7) to mildly acidic in the immature vesicles (pH ~6) to mostly acidic in mature secretory granules (pH ~5) [362]. These differences in pH are very tightly regulated since a difference of even pH<0.5 can affect the ability of a peptidase to properly process a propeptide [363]. Examples of secretory pathway peptidases whose functions are tightly regulated by the pH of their vesicular environment include furin, PC1/3, PC2, and the carboxypeptidases D and E. Early endopeptidic cleavages at the basic sites of many proneuropeptides are performed by furin, which localizes to the Golgi apparatus, while cleavages in later parts of the secretory pathway, including in immature and mature secretory vesicles, are performed by the PCs. Furin is expressed in virtually all cell types, while PC1/3 and PC2 are expressed exclusively in neuroendocrine cells [364]. Furin has a neutral pH optimum, which restricts its activity to the TGN, while PC1/3 and PC2 have acidic pH optima making them active in the mature secretory vesicles [364]. The metallocarboxypeptidases are also regulated by their pH optima with carboxypeptidase D (CPD) being active earlier in the secretory pathway at a more neutral pH [365, 366] while carboxypeptidase E (CPE) functions later in the secretory pathway at a more acidic pH [367]. The endopeptidic cleavage step, at intra-peptidic basic sites, dictates which specific peptides are available for trimming by the carboxypeptidases. This availability, as well as the localization of these peptidases in concurrence with their optimum cleavage conditions help to determine the specific substrates of each enzyme.

Many peptides require further modifications before becoming biologically active. These include phosphorylation, acetylation of the N-terminus, and amidation of the carboxyl-terminus; the latter is carried out by peptidylglycine α-amidating monooxygenase (PAM) [368, 369]. An important point to note is that the timing and subcellular localization of the endoproteolytic cleavage steps play an important role in regulating the processing and packaging of neuropeptides. For example, cleavage of the precursor in the Golgi can lead to independent sorting of the various neuropeptide fragments, while cleavage after packaging into the dense core vesicles ensures the colocalization and co-release of neuropeptides [370]. Once neuropeptides have been processed to maturity, they are stored in mature secretory granules or LDCVs from where they undergo release upon activity dependent cell stimulation [353].

While the classical pathway of peptide processing has been well-established, it has been suggested that in some cases a second independent pathway of peptide processing occurs in the secretory system of neuroendocrine cells [371–373]. This putative pathway involves processing by cathepsin L, at either the middle or the N-terminal side of paired basic residues; these basic residues are subsequently removed by an Arg/Lys aminopeptidase to generate a mature peptide [374]. However, further studies are needed to characterize and ascertain the physiological relevance of cathepsin L in neuropeptide processing. In the following sections we describe examples of precursors that are processed by the classical pathway (i.e., involving PCs).

4.2.1 DIFFERENTIAL PROCESSING: PROOPIOMELANOCORTIN

POMC processing yields several neuropeptides with differing physiological roles [375]. Processing of POMC by PC1/3 yields proadrenocorticotrophic hormone (proACTH) and β-lipotropin [375]. ProACTH is further cleaved by PC1/3 to yield N-terminal peptide, joining peptide and ACTH [375]. PC2 can cleave ACTH to yield ACTH 1-17 and corticotrophin-like intermediate lobe peptide [375] and further processing of ACTH 1-17 by CPE, PAM and N-acetyltransferase gives rise to α-melanocyte stimulating hormone (α-MSH) [375]. β-lipotropin can be proteolytically cleaved by PC2 into γ-lipotropin and β-endorphin [375]. γ-lipotropin is cleaved by PC2 into β-MSH [375]. N-terminal peptide is processed by PC2 into γ-MSH [375]. Interestingly, the neuropeptides obtained from processing of POMC depend on the enzymes expressed in a given tissue or cell type. For example, the anterior pituitary gland lacks PC2 and the major neuropeptide obtained from POMC processing is ACTH. In contrast, the processing of POMC is more extensive in the hypothalamus and intermediate pituitary and results in the formation of α-MSH. Major neuropeptides derived from tissue-specific POMC cleavage are shown in Figure 4.1.

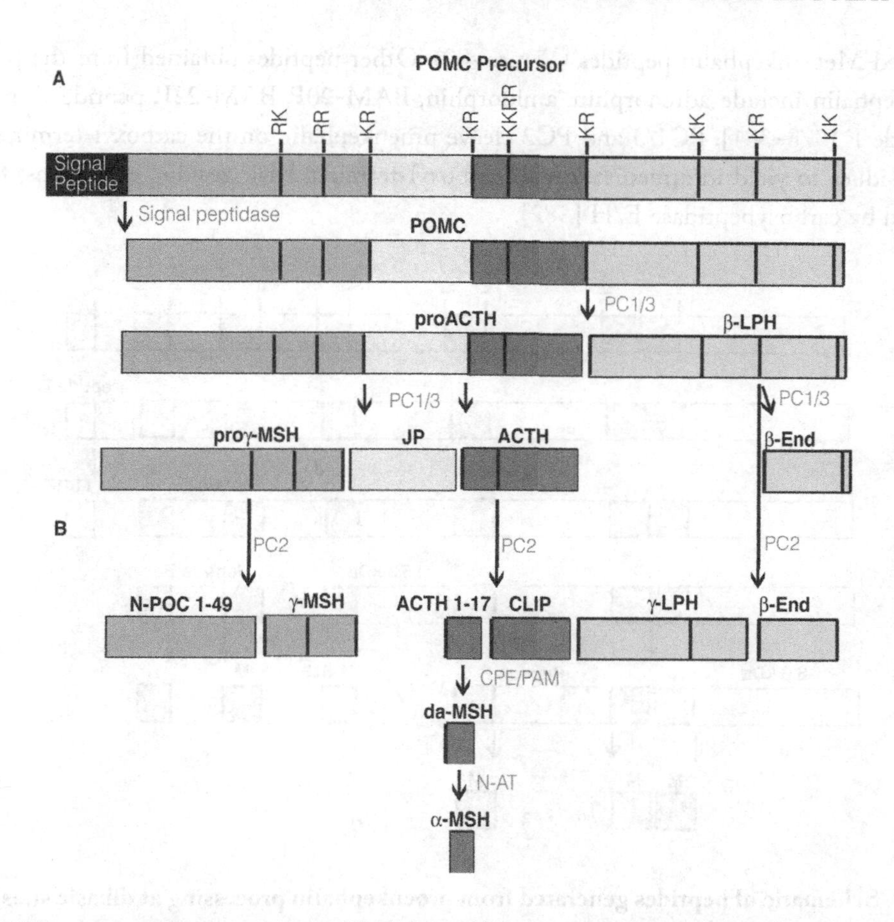

Figure 4.1. **Tissue-specific processing of POMC.** (**A**) In the anterior pituitary which lacks PC2, the POMC precursor is processed by PC1/3 into proγ-MSH, joining peptide (JP), ACTH and β-lipotropin (β-LPH). Limited proteolysis of β-LPH by PC1/3 yields small amounts of β-endorphin (β-End). (**B**) In tissues containing both PC1/3 and PC2 (e.g., hypothalamus, intermediate pituitary) the POMC precursor is further processed into N-terminal pro-opiocortin (N-POC 1-49), γ-MSH, corticotrophin-like intermediate lobe peptide (CLIP), ACTH 1-17, γ-LPH and β-End. ACTH 1-17 is processed by carboxypeptidase E (CPE) and peptidylglycine α-amidating monooxygenase (PAM) into desacetyl-α-MSH (da-MSH) which is further converted by N-acetyltransferase (N-AT) into mature α-MSH.

4.2.2 DIFFERENTIAL PROCESSING: PROENKEPHALIN

Proenkephalin, best known as the precursor of the opioid peptides known collectively as enkephalins, is widely distributed in the brain [376, 377]. Processing of proenkephalin yields multiple copies of the same peptide: one copy of Leu-enkephalin, four copies of Met-enkephalin and two copies of extended Met-enkephalin peptides (Figure 4.2). Other peptides obtained from the processing of proenkephalin include adrenorphin, amidorphin, BAM-20P, BAM-22P, peptide B, peptide E and peptide F [378–381]. PC1/3 and PC2 cleave proenkephalin on the carboxyl-terminal side of dibasic residues to yield intermediates with carboxyl-terminal basic residue extensions; the latter are cleaved by carboxypeptidase E/H [382].

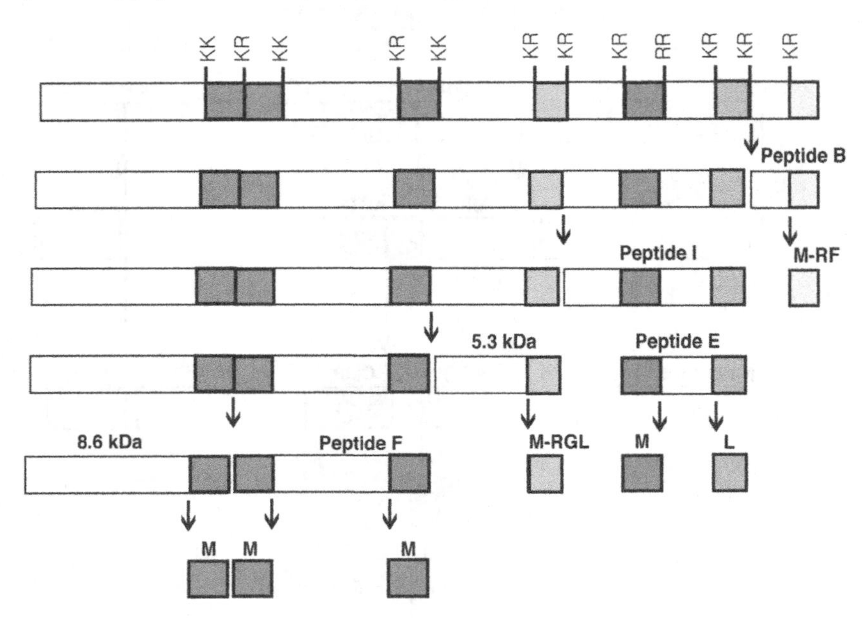

Figure 4.2. **Schematic of peptides generated from proenkephalin processing at dibasic sites.** L, Leu-enkephalin; M, Met-enkephalin; M-RF, Met-enkephalin RF; M-RGL, Met-enkephalin RGL.

4.3 NEUROPEPTIDE SORTING

Neuropeptides are eventually stored in LDCVs prior to secretion; however, not all neuropeptides produced by a given neuron will be found in each of these vesicles [383]. There are cases where two peptides derived from the same precursor are sorted to separate vesicles. A case in point is the *Aplysia Californica* egg laying peptide [384]. Neuropeptide sorting events help to differentiate vesicular cargoes with different functions expressed in the same neurons [385]. While this process is not completely understood, there are two models that attempt to explain this phenomenon. One model suggests a membrane-based sorting of the peptides, where a sorting motif in a given peptide is recognized by a membrane-associated element, either protein or lipid in nature. This would direct the peptide into a particular vesicle [386, 387]. The other model suggests that neuropeptides aggregate passively, interacting with one another to form complexes that ultimately self-segregate

[388]. Peptide aggregation could be triggered following a decrease in the pH of the TGN, by the initial endoproteolytic cleavage step or by the presence of accessory proteins in the TGN [370]. As it currently stands, further studies are required to better elucidate how neuropeptides are sorted into secretory vesicles.

A class of proteins that could play a role in the sorting of neuropeptides comprises the acidic glycoproteins known as granins. The granin family includes chromogranin A, chromogranin B, secretogranin II (also known as chromogranin C), secretogranin III (or 1B1075), secretogranin IV (or HISL-19), secretogranin V (or 7B2), and secretogranin VI (or NESP55) [389]. Studies indicate that, in addition to contributing to the formation of secretory vesicles when immature vesicles bud from the TGN, the granins may play a role in the formation of neuropeptide storage complexes. It is also thought that granins may function as chaperones in sorting regulated secretory neuropeptides [389]. Granins have multiple recognition sites for PC1 and PC2, indicating that they may be proteolytically processed in a specific manner by these enzymes in the regulated secretory pathway. Interestingly, a peptide from the C-terminal region of secretogranin V (or 7B2) inhibits the activity of PC2, while the N-terminal domain of secretogranin V functions as a chaperone for this enzyme in the ER and may be required for full enzyme activity in the early secretory pathway [390]. This is an example of one mechanism by which a neuropeptide can actually exert control over the differential cleavage of its co-secreted peptides by activating their respective endopeptidases.

While there are examples of specific sorting of neuropeptides to distinct vesicular populations, there are also many cases in which peptides from different precursor proteins are co-stored in vesicles awaiting release. For example, substance P has been shown to co-localize with calcitonin gene related peptide (CGRP) in secretory vesicles of peripheral nerves [391]. There is also evidence of substance P being co-stored with galanin [392], somatostatin [393], and BDNF [394] in the same LDCV in sensory afferent terminals. The functional cooperativity of neuropeptides co-localized to the same vesicle has yet to be fully elucidated, but current knowledge holds that they are, at minimum, released in response to the same stimuli.

4.4 NEUROPEPTIDE SECRETION

The secretion of neuropeptides from secretory vesicles is a tightly regulated process. Neuropeptides can be secreted either via the regulated secretory pathway or via the constitutive secretory pathway. There are fundamental differences between the regulated and the constitutive secretory pathways that distinguish the character of the peptides they release. A distinguishing feature of the constitutive secretory pathway is the continuous, unstimulated secretion of proteins or peptides. In this case, small secretory vesicles are formed at the TGN, and constantly bud off, releasing their contents at the plasma membrane without ever being stored in LDCVs. The rate of secretion depends only upon how fast the proteins or peptides are generated in the endoplasmic reticulum [395, 396].

Secretion via the regulated secretory pathway differs mostly with respect to peptide storage. Before secretion, peptides accumulate in secretory vesicles until they form a dense core, characteristically distinguishable by electron microscopy. While synaptic vesicles containing the classic, small molecule neurotransmitters, such as acetylcholine and norepinephrine, are targeted directly into the active zones of the synapses, neuropeptide release from LDCVs occurs at axonal terminals away from the synapse, and at the neuronal soma [397, 398].

In the case of POMC, studies suggest that the unprocessed precursor may be secreted via the constitutive secretory pathway while neuropeptides generated from its processing are secreted through the regulated secretory pathway [375]. It has been proposed that the initial cleavage of POMC into proACTH and β-lipotropin may be rate limiting, thereby determining the amount of precursor processed in the secretory pathway [375]. Unprocessed POMC is then constitutively secreted, while proACTH and β-lipotropin are sorted into secretory vesicles via an as yet poorly understood mechanism. It is thought that the sorting and retention of POMC-derived peptides into the regulated secretory pathway involves a number of sorting motifs, including an N-terminal amphipathic loop stabilized by a disulfide bridge acting as a sorting signal motif that binds to CPE as a putative sorting receptor [375].

Neuropeptides are also regulated at the level of secretion. The peptides are produced and then stored in secretory granules of neuroendocrine cells, where they await release [399]. Secretion of neuropeptides from secretory granules of a given neuron can be induced by action potentials that lead to neuronal depolarization. The latter can be induced by neighboring neurons either through impinging excitatory currents, or via the release of ligands that bind to cell surface receptors of the neuron in question [400]. The stored pool of neuropeptide-containing vesicles allows for the increased release of neuropeptide without needing to translate and process more peptides. The basal secretion of neuropeptides is very low—roughly 0.5% of the cell peptide content per hour. This indicates that the vast majority of peptides are kept in reserve for secretion upon stimulation [353].

Another level of neuropeptide regulation is provided by the particular complement of processing enzymes present in the cell's regulated secretory pathway. For example, POMC is differentially cleaved in the anterior pituitary gland to produce adrenocorticotropic hormone (ACTH), while it can be alternatively cleaved in anorexigenic neurons of the arcuate nucleus to produce α-MSH [401, 402] (Fig. 4.1).

4.5 NEUROPEPTIDE DEGRADATION

Following secretion, neuropeptides exert their biological effects by binding to cell surface receptors. Commonly, the receptors for neuropeptides are GPCRs (see Chapters 2 and 3). The major mechanism by which neuropeptide signaling is terminated, or the extracellular concentration of the neuropeptides is adjusted, involves peptide degradation. Peptidases anchored to the cell membrane with their active site facing the extracellular space have been implicated in the degradation of neuropeptides in the central nervous system. These include endopeptidase 24.11 (also known as enkephalinase or neprilysin), angiotensin-converting enzyme, aminopeptidase M (also known as aminopeptidase N), pyroglutamyl aminopeptidase II, dipeptidyl peptidase IV [403] (Table 4.2). *In vitro* studies show that prolyl endopeptidase, calpain, endopeptidase 24.15, endopeptidase 24.16, aminopeptidases A and B and puromycin sensitive aminopeptidase may also be involved in neuropeptide degradation [403].

Table 4.2. Neuropeptide degrading enzymes

Enzyme	Distribution	Cleavage site	Substrates	References
Endopeptidase 24.11	Nervous system	Amino terminus	Enkephalins; substance P; neurokinins; neurotensin; endothelins	[714]
Angiotensin-converting enzyme (ACE)	Brain; spinal cord; lungs; kidneys; intestine; blood vessels; placenta; lymph nodes; retina	Carboxyl terminus	Angiotensin I; bradykinin; neurotensin; substance P; luteinizing hormone releasing hormone; dynorphins; enkephalins	[714]
Aminopeptidase M	Central nervous system; intestine; placenta; lymph nodes; lungs; liver; blood vessels	Amino terminus	Enkephalins; neurokinins; somatostatin; dynorphins	[715]
Pyroglutamyl aminopeptidase II	Brain; liver; spinal cord; kidneys; adrenal gland	Pyroglutamate-histidine bond	Thyrotropin releasing hormone	[716]
Dipeptidyl peptidase IV	Intestines; placenta; lymph nodes; liver; blood vessels; pancreas; kidneys	X-Pro-Z and X-Ala-Z (X=non-specific amino acid; Z= cannot be Pro or hydroxyPro)	Substance P	[717]
Endothelin-converting enzyme 1	Cardiovascular, reproductive & endocrine systems	Amino side of hydrophobic residues	Endothelin; substance P	[405]

Endopeptidase 24.11 was first detected because of its ability to degrade enkephalins, and is likely the major neuropeptide degrading enzyme. This enzyme is widely distributed throughout the central nervous system, and is present predominantly in neuronal cells at both pre- and post-synaptic sites [403] (Table 4.2). The enzyme is also present in the spinal cord, intestine, kidneys and a number of other tissues [403]. Endopeptidase 24.11 preferentially cleaves substrates on the amino side of hydrophobic residues (e.g., Met, Leu, Phe) [403]. In addition to enkephalins, other neuropeptides cleaved by endopeptidase 24.11 include substance P, neurokinins, neurotensin and endothelins [403].

Angiotensin-converting enzyme, an enzyme that converts angiotensin I to angiotensin II and inactivates bradykinin, has been shown to degrade a number of neuropeptides including neurotensin, substance P, luteinizing hormone releasing hormone, dynorphins and enkephalins [403]. This enzyme has been purified from brain and spinal cord as well as lungs, kidneys, intestine, blood vessels, placenta, lymph nodes and retina. The enzyme has a broad substrate specificity and its action generates C-terminal dipeptides (Table 4.2) [403].

Aminopeptidase M is a major membrane-associated aminopeptidase that can also degrade enkephalins. This enzyme is found at the pre- and post-synapse as well as in non-neuronal cells such as astrocytes. It is also found outside the CNS in the intestine, placenta, lymph nodes, lungs liver and blood vessels [403]. In addition to enkephalins, aminopeptidase M can cleave neurokinins, somatostatin and dynorphin A 1-17 [403]. Aminopeptidase M is an exopeptidase that cleaves N-terminal amino acid residues from peptides. The enzyme is inactive toward proline containing N-terminal sequences or acetylated N-terminal residues (Table 4.2) [403].

Pyroglutamyl aminopeptidase II is a membrane-associated metallopeptidase that has been shown to specifically hydrolyze the pyroglutamate-histidine bond of thyrotropin-releasing hormone [403]. This enzyme is found in the brain, liver, spinal cord, kidney and adrenal gland (Table 4.2) [403].

Endothelin-converting enzyme 1 (ECE-1) functions not only in the generation of endothelin peptides, but is also involved in neuropeptide degradation. ECE-1, a zinc metalloendopeptidase with an amino acid sequence related to neprilysin, generates active peptides at the cell surface. This peptidase can hydrolyse many substrates *in vitro*, including neurotensin, substance P, and bradykinin, with similar efficiency to known *in vivo* substrates [404]. ECE-1 cleaves substrates at the amino side of hydrophobic residues. ECE-1 has been shown to regulate tachykinin receptor activation by degrading substance P bound to the receptor in the endosomal compartment [405].

Together, the action of a neuropeptide is regulated at multiple levels, from transcriptional regulation of the precursor protein to differential processing, secretion and degradation. The major function of the neuropeptide *in vivo* is to activate its cognate receptor system. This is described in detail in Chapter 5.

CHAPTER 5

Neuropeptide Receptors

5.1 INTRODUCTION

The activity and function of neuronal circuits in the brain is modulated by peptides present in the milieu surrounding individual neurons. Endogenous neuropeptides generated as described in Chapter 4 exert their effects by binding to cell-surface receptors, which in many cases are GPCRs. In addition, peptides such as nerve growth factor and brain-derived growth factors that are generated in the brain as well as peptides generated in the periphery, such as leptin, and insulin can exert some of their physiological effects by binding to receptors in the brain which in the case of these peptides are enzyme-linked receptors [406].

In the following sections we describe some important peptide-receptor systems, how manipulations of either peptide or receptor levels affect function, and their involvement in pathophysiological conditions.

5.2 NEUROPEPTIDES AND THEIR RECEPTORS

5.2.1 OPIOID PEPTIDES & OPIOID RECEPTORS

Opioid peptides

Opioid peptides comprise of three well-characterized functionally active neuropeptide groups: enkephalins, endorphins and dynorphins. These peptides are derived from the processing of large precursors termed POMC, proenkephalin and prodynorphin [407]. Processing of POMC yields β-endorphin as well as non-opioid peptides like ACTH, β-lipotrophin, α- and γ-MSH [407] (see Chapter 4). Processing of proenkephalin yields a number of opioid peptides including four Met-enkephalin peptides, two carboxyl extended Met-enkephalin peptides, and one Leu-enkephalin peptide [407]. In addition, differential processing of proenkephalin can yield larger peptide intermediates such as peptides E and F [407] (see Chapter 4). Processing of prodynorphin also yields a number of opioid peptides including Leu-enkephalin, dynorphins, β-neoendorphin and leumorphin. Like all opioid peptides, dynorphins, β-neoendorphin and leumorphin contain the enkephalin sequence at the amino terminus, indicating that the cleavage of one functional opioid peptide can give rise to another opioid peptide with a different function and receptor affinity [407]. The following sections focus on the three major types of opioid peptides and the receptors through which they exert their physiologic responses.

Endorphins

Endorphins are a class of opioid peptides involved in antinociception. In addition to antinociception, endorphins play a major role in the regulation of the hypothalamic-pituitary-gonadal axis activity, the cellular effects of exercise, in stress-induced analgesia as well as in mood-enhancing and anxiolytic effects [408]. These peptides are derived from the proteolytic processing of POMC, with the major analgesic peptide being the 31 amino acid β-endorphin peptide [409, 410]. β-endorphin is generated primarily in the hypothalamus and the pituitary gland [409, 410]. Recent studies show that β-endorphin can also be synthesized in cells of the immune system as part of the inflammatory response [410]. This peptide is released from the hypothalamus and pituitary in response to traumatic injury as well as exercise and other stimuli. The action of β-endorphin in pain perception is inhibitory in nature, inhibiting neuronal firing to peripheral somatosensory fibers, an effect that is abrogated by the addition of naloxone [411]. Higher levels of the peptide correlate with reduction of observed pain in humans during oral surgery [412].

Mice lacking β-endorphin also display alterations in food intake consistent with a role for endorphin in the incentive-motivation and reward associated with food acquisition [413]. In addition, mice lacking β-endorphin do not develop opioid stress-induced analgesia [414]. Also indicative of the involvement of β-endorphin in the reward pathway, a recent study indicated that β-endorphin is required for the rewarding actions of acute cocaine [415].

β-endorphin is a potent agonist of μ-opioid receptors, and is up to 80 times more potent than the exogenous agonist morphine [409]. These receptors are described below under Opioid receptors.

Enkephalins

Enkephalins comprise a major class of neuropeptides involved in analgesia and pain perception. There are two major enkephalin peptides that are identical in sequence except for their final amino acid, Met-enkephalin (Tyr-Gly-Gly-Phe-Met) and Leu-enkephalin (Tyr-Gly-Gly-Phe-Leu) [381]. Enkephalins are found in brain and spinal cord regions involved in the processing of pain impulses (e.g., lower brain stem, periaqueductal grey region, substantia gelatinosa) [416]. High densities of enkephalinergic neurons are also observed in amygdaloid nuclei, hypothalamus, and the globus pallidus, brain regions involved in the regulation of behavior and emotional control, secretion of releasing hormones and motor activity [416]. Enkephalins are also found in the intestinal tract where they play a role in peristalsis and anti-diarrhetic action (Table 5.1) [416].

Table 5.1. Neuropeptides receptors

Receptor	Subtypes	Neuropeptide ligands	Distribution	G protein coupling	Receptor knockdown effects	References
Opioids	μ, δ, κ	Enkephalin; β-endorphin; dynorphin	Central & peripheral nervous system; intestinal tract	Gαi/o; adenylyl cyclases; Ca^{+2} channels	No response to drugs of abuse; reduced feeding; altered maternal attachment; depression; anxiety	[50, 349, 407, 409, 416, 430, 432, 718, 719]
Melanocortin	MC1R; MC2R; MC3R; MC4R; MC5R	α-MSH; AgRP	Brain; neuroendocrine tissues; heart; lungs; fat cells; skin	Gαs; adenylyl cyclases;	Obesity; hyperphagia; albinism	[385, 445, 446, 452]
Neuropeptide Y	Y1; Y2; Y3; Y4; Y5	NPY; Peptide YY	Broad distribution in brain	Gαi/o; adenylyl cyclases; Ca^{+2} channels	Knockdown has no effect on food intake or body weight	[456, 460, 470, 472]
Neurokinin	NK1; NK2; NK3	Substance P; tachykinin peptides	NK1 distributed in central & peripheral nervous tissue	Gαq; PLC	↓capsaicin response; ↓lung inflammation	[486–488, 493]
Vasopressin	V1A; V1B; V2	Vasopressin	Brain; liver; kidneys; vascular smooth muscle	Gαq; PLC; Gαs; adenylyl cyclases	defective pair bonding; impaired diuresis	[522–524, 720]
Oxytocin	Alleles with differing effectiveness	Oxytocin	Brain; neuroendocrine tissue; mammary & uterine epithelium	Gαq; PLC	abnormal emotional & social behavior; ↑aggression; impaired nurturing	[504–506, 512]

Endocan-nabinoid	CB_1R; CB_2R	Hemopressins	Brain; adi-pose tissue; immune cells	Gαi/o; adenylyl cyclases;	leaness; ↓re-ward response to drugs of abuse	[527, 721–724]

Studies with mice lacking proenkephalin show that they exhibit increased pain responses in the hot plate and tail flick tests but decreased pain perception following formalin administration [417]. In addition, these mice exhibit attenuation of food seeking behavior [413], reduced locomotor activity in the open-field assay and higher levels of anxiety than their wild-type controls [417].

Enkephalins are rapidly degraded and have a turnover rate of seconds to minutes [416]. A major catabolic step involves cleavage of the Tyr1-Gly2 bond by a membrane associated aminopeptidase [416]. Several enzymes are thought to be involved in enkephalin catabolism include enkephalinases, non-specific aminopeptidases, carboxypeptidases, cathepsin C and angiotensin-converting enzyme [416].

Enkephalins exert their biological effects through the activation of μ- or δ-opioid receptors [418]. These receptors are described in greater detail below under Opioid receptors.

Dynorphins

Dynorphins are another class of neuropeptides that regulate pain perception. The proteolytic processing of prodynorphin yields multiple peptide products including Dyn A1–17 (Dyn A), Dyn A1–8, Dyn A1–13, Dyn B1–13 (Dyn B/rimorphin), Dyn B1–29 (leumorphin), Dyn A/B1–32 (big DYN) and α/β-neo-endorphin, all of which have biological effects [419]. These peptides are generated by the processing of prodynorphin by PC1/3, PC2 and carboxypeptidase E [420]. Within the brain, the dynorphin peptides are most concentrated in the hypothalamus, medulla, pons, and midbrain [421]. Dynorphins play a role in a number of physiological processes including pain attenuation, neuroendocrine regulation, motor activity, cardiovascular function, respiration, temperature regulation, feeding, circadian rhythms, depression, addiction, and stress responses [422-426]. The activity of these highly potent peptides are partially blocked by the non-selective opioid receptor antagonist naloxone [427]. This suggests that dynorphins exert at least some of their effects via a system differing from that of morphine or the other opioid peptides. This is supported by studies showing that the development of tolerance to morphine-mediated analgesia has no effect on dynorphin-mediated analgesia [427].

Mice lacking prodynorphin exhibit increased pain responses in the hot plate and tail flick tests [417]. Interestingly while spinal nerve ligation leads to sustained neuropathic pain in wild-type animals, pain responses in mice lacking prodynorphin returned to basal levels 10 days after nerve ligation [428]. This, together with the observation that lumbar dynorphin levels are increased in the wild-type animals exhibiting neuropathic pain, suggests that dynorphins may be involved in the maintenance of neuropathic pain [428]. More recently, studies showed that mice lacking prodynorphin exhibit increases in cue dependent fear conditioning, a delay in the extinction of contextual condition of fear paradigms suggesting that dynorphin peptides may play a role in the regulation of fear memory [429].

Early characterization of the function of the dynorphin peptides showed that they are powerful modulators of opioid receptors, with highly potent effects when compared to enkephalins and morphine [430]. Although dynorphins bind to all three subtypes of the opioid receptor they exhibit a greater selectivity toward κ opioid receptors [422]. These receptors are described below under Opioid receptors.

Opioid receptors

Opioid peptides exert their physiologic effects by activating family A GPCRs termed opioid receptors. The three major types of opioid receptors are μ-, δ- and κ-opioid receptors. While endorphins exhibit selectivity for μ-opioid receptors, enkephalins exhibit selectivity to μ- and δ-opioid receptors and dynorphins to κ-opioid receptors (Table 5.1) [431]. The three types of opioid receptors share similar structural elements with Family A GPCRs (see Chapter 2). The different opioid receptor types share roughly 60% identity, with the highest sequence homology observed in the TM and intracellular loop regions. Conversely, the N- and C-terminal regions of opioid receptor types show the greatest divergence in amino acid sequence [432]. Opioid receptors are the products of three separate genes, *OPRM1*, *OPRD1*, and *OPRK1* which are located on different chromosomes. In addition, several isoforms of opioid receptors have been identified in human tissue, and are thought to arise from alternative splicing [433].

The μ-, δ- and κ-opioid receptors are expressed throughout the central nervous system, the peripheral nervous system, as well as the intestinal tract. In the brain opioid receptors are detected in regions associated with nociceptive, sensory, neuroendocrine, homeostatic and behavioral functions [434]. In general, the tissue distribution of opioid receptors matches that of the endogenous opioid peptides. The three subtypes of opioid receptors show unique but overlapping patterns of distribution in the brain. In general μ-receptors are abundantly present in thalamic and epithalamic areas, δ receptors in cortical and subcortical structures and κ-receptors in thalamic and hypothalamic regions. In addition, all three opioid receptor subtypes are present in the hippocampus, nucleus accumbens and caudate putamen [435].

Opioid receptors couple to Gαi/o proteins and their activation leads to inhibition of adenylyl cyclase activity as well as of calcium channels [436] (Chapter 3). These receptors can also activate PLC, GIRKs, and the extracellular regulated MAPKs, ERK-1 and ERK-2. By increasing K^+ conductance following agonist binding, activation of these receptors leads to membrane hyperpolarization, and a subsequent reduction in the docking and release of classical neurotransmitters from presynaptic neuronal terminals [436].

The functions of specific opioid receptors have been elucidated by the generation of mice lacking specific receptor types. Mice lacking μ-opioid receptors do not exhibit morphine-mediated analgesia, morphine-mediated hyperlocomotion, respiratory depression, inhibition of gastrointestinal mobility and immunosuppression. In addition, the rewarding effects of morphine and of drugs of abuse such as marijuana, alcohol and nicotine (as measured by conditioned place preference (CPP) or self-administration paradigms) are absent in mice lacking μ-opioid receptors [417, 437–439]. Taken together, these studies show that μ-opioid receptors are the primary target for the actions of morphine *in vivo*. δ-opioid receptors appear to be involved in the regulation of emotional responses, since mice lacking these receptors exhibit increased levels of anxiety and a depressive-like phenotype [417, 440]. Studies with mice lacking κ-opioid receptors show that this receptor mediates the analgesic effects of dynorphins and contributes to the dysphoric effects of opioids and cannabinoids [417, 441–443]. Mice lacking either μ-, δ-, or κ-opioid receptors exhibit enhanced sensitivity to pain, indicating that these receptors play a role in inhibition of nociceptive responses [417].

Together, these studies imply a role for μ-receptors in the modulation of mechanical-, chemical- and supraspinally-mediated thermal nociception, δ-receptors in reducing hyperalgesia during inflammatory and neuropathic pain, and κ-receptors in thermal nociception and visceral pain [417].

The activation of opioid receptors leads to a broad range of physiological effects. Even taking into account the diversity of receptor types, in conjunction with their differential localization, it is nearly impossible to account for the many, varied physiological effects they mediate. Various factors could mediate these effects, including posttranslational modifications, alternative splicing, scaffolding effects and receptor heterodimerization [433]. The effect of heteromerization on opioid receptor function is described in Chapter 3.

5.2.2 α-MSH, AGRP AND MELANOCORTIN RECEPTORS

The melanocortin system is activated by α-MSH, γ-MSH and ACTH, all of which are generated from the processing of POMC (Figure 4.1). This system is inactivated by the endogenous antagonist, AgRP [444, 445]. The melanocortin system has been implicated in a wide array of physiological processes such as pigmentation, steroidogenesis, energy homeostasis, feeding, inflammation, immunomodulation, analgesia, temperature control and cardiovascular regulation [445]. In the following sections we describe the involvement of α-MSH, AgRP and melanocortin receptors in the regulation of feeding.

α-MSH

The main food intake regulating peptide derived from POMC is α-MSH, a 13 amino acid neuropeptide. α-MSH is expressed in the pituitary, hypothalamic arcuate nucleus and in the nucleus tractus solitarius of the brainstem where it has a crucial role in the regulation of metabolic functions [446]. The neuroendocrine tissue where the peptide is generated has important effects on its function since α-MSH generated in the arcuate nucleus of the hypothalamus has effects on food intake and bodyweight, whereas α-MSH produced by the pituitary functions primarily in the regulation of melanin production [447, 448]. The production and release of α-MSH from POMC containing neurons in the arcuate nucleus is promoted by leptin [446]. The released α-MSH binds to and activates melanocortin MC3 and 4 receptors, leading to suppression of food intake and an increase in energy expenditure [448]. The activity of POMC-containing neurons is modulated by neighboring neurons that produce NPY, AgRP and GABA. Leptin has complex interactions activating POMC neurons in the arcuate nucleus while simultaneously suppressing the activity of NPY/AgRP containing neurons. This prevents the release of AgRP and consequently the ability of the latter to block the binding of α-MSH to melanocortin receptors [446]. This system is also regulated by ligand degradation by prolylcarboxypeptidase, which removes the C-terminal valine residue from α-MSH leading to its inactivation.

AgRP

AgRP is biosynthesized as a 132 amino acid peptide (and is not derived from the POMC precursor). AgRP antagonizes the action of α-MSH [385]. The interaction between these two peptides results in a dynamic control of food intake and metabolism. The name AgRP stands for "agouti related peptide" and stems from the fact that agouti, a coat color gene in mice [385], is generated only in the arcuate nucleus of the hypothalamus where it localizes to NPY containing neurons [449]. Although the mRNA levels of AgRP and of NPY are upregulated in response to fasting in the mouse brain and the effects of acute peptide administration into the ventricle are similar, the duration of the effects of the two peptides differ. While NPY has potent, short-lived effects in increasing food intake, AgRP can induce an increase in food intake for up to a week [450]. The involvement of AgRP in promoting food intake and energy expenditure is supported by studies showing that the central administration of this peptide or peptide overexpression leads to hyperphagia and obesity [449, 451].

Melanocortin receptors

Both α-MSH and AgRP exert their biological effects by binding to melanocortin (MC) receptors. While α-MSH is an endogenous agonist of melanocortin receptors, AgRP functions as an endogenous antagonist. MC receptors belong to the family A of GPCRs and are coupled to Gαs proteins. Thus receptor activation leads to increases in intracellular cyclic AMP levels and consequently PKA activation (see Chapter 3). Five different subtypes of MC receptors, MC1R to MC5R, have been identified that are the products of five separate genes [452]. All five MC receptor subtypes can be activated by neuropeptides derived from the processing of POMC. ACTH is a full agonist of MC1R, MC2R and MC3R and a partial agonist of MC4R and MC5R. α-MSH and β-MSH are full agonists at MC1R, MC3R, MC4R and MC5R. γ-MSH is a full agonist of MC3R and a partial agonist of MC1R, MC4R and MC5R [452].

MC1R is present mainly in the skin and hair follicles where it plays a role in skin, hair and coat pigmentation. MC2R is expressed in the adrenal glands and its activation by ACTH leads to the production of glucocorticoids. MC3R and MC4R are widely expressed in the hypothalamus and play a role in food intake and energy expenditure. MC5R is expressed in sweat glands cells where it promotes the production of an oily/waxy material called sebum that lubricates and water-proofs the skin and hair of mammals [453].

Although both MC3R and MC4R play a role in food intake and energy metabolism, knockdown studies indicate that their physiological roles are different. Targeted deletion of MC3R in mice results in adiposity although their food intake is reduced suggesting that the mice preferentially store nutrients as fat. In addition, these mice exhibit a reduction in energy expenditure. Targeted deletion of MC4R leads to maturity onset of severe obesity, hyperphagia, hyperinsulinemia and hyperglycemia (Table 5.1). These mice gain weight faster than their wild-type littermate controls while consuming the same amount of food [449]. Rescue of MC4R expression in MC4R knockout mice indicates that this receptor expressed in the PVN is responsible for food intake while the receptor expressed in other brain regions is responsible for energy expenditure [449].

5.2.3 NPY AND NPY RECEPTORS

NPY

NPY, a 36 residue neuropeptide, is a member of a peptide family that also includes two gut hormones, peptide YY and pancreatic polypeptide [454, 455]. NPY is named for having a tyrosine residue at both the N- and C-terminus of the peptide, and is one of the most evolutionary conserved peptides. It is synthesized from the proNPY precursor that is proteolytically cleaved at dibasic residues to generate NPY and the C-terminal peptide of NPY (CPON) [456,457]. The C-terminus of NPY is amidated, which is required for its biological activity [456].

NPY is abundantly expressed in the brain particularly in hypothalamus, amygdala, hippocampus, nucleus of the solitary tract, locus coeruleus, nucleus acumbens and cerebral cortex [458]. It is also abundantly expressed in peripheral tissues including adrenal medulla, liver, heart, spleen and endothelial cells. In the brain, NPY colocalizes with norepinephrine, GABA, and somatostatin in agouti-related protein-containing neurons in the arcuate nucleus [459,460].

NPY plays a role in a number of physiological processes including stress response, food intake, control of energy balance, sleep regulation, inflammatory processes, cardiovascular regulation, memory, nociception, seizures and tissue growth and remodeling [461–463]. Not much is known about the functional role of CPON [456].

Studies show that intracerebroventricular administration of NPY results in robust increases in food intake, leading to an increase in body weight after sustained NPY administration [464–466]. This increase in body weight is due not only to NPY-induced acute hyperphagia, but also to changes in the fundamental metabolic parameters of the treated animals [467]. Knockdown of the NPY gene results in several interesting phenotypes, including altered sensitivity to ethanol and spatial memory as well as anxiety-like behavior. Interestingly, knockdown of the NPY gene does not lead to significant changes in feeding or bodyweight, despite the well-characterized role the peptide plays in these processes, likely due to the highly redundant nature of feeding control by neuropeptides [468, 469].

NPY receptors

NPY exerts its physiological effects by activating family A GPCRs termed NPY receptors [470]. The major functional subtypes of NPY receptors (NPY1, NPY2, NPY4, and NPY5) are individually encoded by different genes [456]. NPY1 receptor mediates vascular and antinociceptive effects, feeding response and ethanol consumption [471]. Activation of this receptor subtype by NPY leads to reduction in anxiety and depression [471, 472]. NPY2 receptor plays a role in the vascular effects of NPY such as angiogenesis and blood pressure regulation, as well as in the regulation of circadian rhythms, bone formation and feeding responses. NPY4 receptor may be involved in regulating reproduction and energy homeostasis while the NPY5 receptor plays a role in feeding behavior, energy expenditure, brain excitability, inhibition of reproductive hormone release and regulation of circadian rhythms [471]. Recent studies suggest that NPY4 and NPY5 receptors are also involved in depression-like behaviors [471, 472].

NPY1, NPY2, NPY4 and NPY5 receptors couple to Gαi/o proteins and their activation therefore results in inhibition of cyclic AMP levels [456] (Chapter 3). Activation of NPY receptors also leads to PTX-sensitive phosphorylation of ERK 1/2 confirming that these receptors couple to Gαi/o proteins [456]. A PKC independent pathway may also be involved in NPY1, NPY2, and NPY4 receptor signaling [456]. In addition, NPY receptors are reported to couple to PLC to induce release of Ca^{2+} from intracellular stores [456].

The NPY receptor subtypes have differing expression patterns and binding affinities for their cognate ligands. These varying affinities and distributions functionally differentiate the receptor subtypes. Thus, NPY1 receptor expression can be detected in thalamic nuclei, hippocampus, caudate nucleus, putamen, nucleus accumbens, amygdaloid nuclei, paraventricular nucleus (PVN) of the hypothalamus, arcuate nucleus, colon, kidney, adrenal gland, pancreatic β cells, placenta and in visceral adipose tissue [473]. The NPY2 receptor, the most prominent receptor subtype present in the central nervous system, is located mainly presynaptically and is involved in the suppression of transmitter release [473]. This receptor subtype can also be found within the hippocampus, hypothalamus and amygdala, as well as in specific nuclei of the brain stem [473]. The NPY4 receptor is found mainly in the brain stem, the dorsal motor nucleus of the vagus nerve, and in the nucleus tractus solitarius, in orexin-containing neurons in the lateral hypothalamic area, and in peripheral tissues such as heart, gastrointestinal tract, skeletal muscle, pancreas, testis and uterus [473]. The NPY5 receptor is present in the amygdala and areas of the brain implicated in the regulation of feeding such as the lateral hypothalamus and overlaps with NPY1 receptor expression in the arcuate nucleus, PVN and suprachiasmatic nucleus of the hypothalamus [473].

NPY receptors are especially dense in areas of the hypothalamus known to regulate food intake and body weight. The density of NPY receptors in the hypothalamus has been shown to decrease with prolonged fasting, probably due to receptor downregulation following exposure to NPY whose secretion is increased during starvation [474]. Studies reveal that the NPY5 receptor subtype may be responsible for the orexigenic effects of NPY [475]. For example, impairment of the NPY5 receptor, either through genetic knockdown or by specific receptor blockade, reduces the increased feeding effects typically observed following a central infusion of NPY [476, 477].

Mice lacking the NPY1 receptor subtype do not exhibit major changes in food intake and body weight although an obese phenotype appears to develop in older mice [468]. These mice exhibit hyperalgesia and mechanical hypersensitivity to acute thermal, cutaneous and visceral chemical pain, increased ethanol consumption, and abrogation of the vasoconstrictive effects of NPY [463, 468, 478]. In the case of mice lacking NPY2 receptors conflicting effects on food intake and body weight were reported. One group reported a decrease in body weight only in male knockout mice. In addition, food intake was unaltered in male mice and increased in female mice lacking this receptor [479]. Another group reported that mice lacking the NPY2 receptor exhibit an increase in body weight, food intake and fat deposition [480]. An interesting phenotype observed with these mice is an increase in trabecular bone volume suggesting that this receptor subtype may play a role in regulation of bone formation [481]. Mice lacking the NPY4 receptor exhibit a decrease in food intake and body weight, a significant decrease in basal blood pressure and a smaller heart compared to control mice [479, 482]. Mice lacking the NPY5 receptor feed and grow normally although they develop late onset obesity (>30 weeks) [483, 484].

5.2.4 TACHYKININS AND NEUROKININ RECEPTORS

Tachykinins, also referred to as neurokinins, comprise three well-characterized functionally active neuropeptides: substance P, neurokinin A, and neurokinin B. Less well-characterized members of this family include neuropeptide K, neuropeptide γ and hemokinin-1 [485]. Tachykinins are characterized by the presence of a common C-terminal sequence Phe-X-Gly-Leu-Met-NH2 where X is either an aromatic or aliphatic amino acid [486]. The tachykinin peptides are generated from the processing of precursors named preprotachykinin A (PPTA), preprotachykinin B (PPTB), and preprotachykinin C [485]. Processing of PPTA generates Substance P, neurokinin A, neuropeptide K and neuropeptide γ, whereas processing of PPTB generates neurokinin B and processing of preprotachykinin C generates hemokinin-1 [485]. These peptides have diverse physiological roles with substance P being the most studied and best characterized peptide-receptor system in the tachykinin family. The following sections focus on substance P and its receptor system.

Substance P

Substance P plays a major role in the modulation of pain pathways, gut motility, emesis, mood, anxiety and stress [487]. Substance P was originally discovered as a peptide influencing intestinal contraction and vasodilation, although it is currently best known for its role in pain signaling from the periphery to the central nervous system [488]. Substance P is expressed throughout the central and peripheral nervous system, including the brain and spinal cord, as well as in neuroendocrine tissues of the gut [487]. Substance P is released from capsaicin-sensitive primary neurons in response to peripheral pain and inflammation. After it is released, substance P can be proteolytically degraded by dipeptidyl-amino peptidase, ECE1, proline endopeptidase, or cathepsin D [405, 486].

Because of its role in nociception substance P has been implicated in fibromyalgia, arthritis, and other inflammatory pain syndromes [489]. Genetic disruption of substance P in mice has confirmed its involvement in pain since these animals exhibit an attenuation of nociceptive pain responses to mechanical, thermal and chemical stimuli [490].

Substance P receptors

Substance P modulates pain responses by binding to and activating neurokinin (NK) receptors in the central nervous system. To date three subtypes of NK receptors, NK1, NK2, and NK3, have been identified that are encoded by *TACR1*, *TACR2*, and *TACR3* genes respectively [486]. NK receptors are GPCRs that signal via Gαq leading to the activation of PLC (see Chapter 3). This, in turn, generates DAG and IP3 that are responsible for downstream signaling through Ca^{2+} release and PKC activation, respectively [491]. Substance P binds to the three subtypes of NK receptors although it exhibits greater affinity for NK_1 receptors (Table 5.1) [492].

NK_1 receptors are present in the brain, spinal cord, vascular smooth muscle, gastrointestinal and genitourinary tract, lungs, thyroid and immune cells [493]. In the nervous system, activation of the NK_1 receptor by substance P transmits pain and stress signals from the peripheral to the central nervous system, and causes smooth muscle contraction in the vascular epithelium [494]. Because of the widespread distribution of the NK_1 receptor, it has been implicated in diseases in many organ systems, including psychiatric disorders such as anxiety and depression in the brain. In the periphery, NK_1 receptors have also been found to be involved in asthma in pulmonary tissue, rheumatoid arthritis in immune inflammation, and gastrointestinal disorders [491,495].

Studies show that mice lacking the NK_1 receptor exhibit a reduced response to capsaicin; sensitivity to the latter can be used to positively identify targets of substance P [496]. Studies with knockout mice support a role for the receptor in pain responses, gastrointestinal distress, and hypoalgesia [490, 497–499]. Thus NK_1 receptor antagonists have therapeutic applications as analgesic drugs, anti-emetics for nausea caused by cancer treatments, and in the treatment of addiction, especially alcoholism (Table 5.1) [487, 500–502].

There is evidence that NK_1 receptors can heterodimerize with other GPCRs leading to modulation of individual receptor dynamics, especially in terms of internalization, recycling or signal resensitization [344, 503].

5.2.5 OXYTOCIN AND OXYTOCIN RECEPTORS

Oxytocin

Oxytocin is a nine amino acid peptide generated by the classical processing of the prooxytocin precursor that is cleaved at dibasic sites to yield oxytocin and neurophysin I [504]. These peptides are produced in the PVN and supraoptic nuclei in neurons which project to the posterior hypothalamus [505]. These peptides are also released from the posterior pituitary [505]. It is thought that neurophysin I serves as a carrier protein for oxytocin from the site of production in the cell body to the site of release at the axonal terminals [504]. Release of oxytocin can be stimulated by a number of peripheral and central stimuli, including sexual stimulation, dilation of the uterus, stress, and nursing. Peripheral oxytocin receptors in the uterus and mammary glands are activated by pituitary oxytocin, resulting in induction of labor and lactation [505]. Oxytocin released from hypothalamic neurons has different physiological effects compared to that released from other neuroendocrine tissues [505]. It is thought that while pituitary gland oxytocin is involved in the physiological role of the peptide including birth and nursing, oxytocin neurons in the brain mediate the behavioral aspects of oxytocin function [505]. Central administration of oxytocin causes gender dependent effects on sexual arousal in rodents [506, 507]. Brain release of oxytocin also has a role in pair bonding in monogamous females rodents, whereas this role is played by vasopressin in male rodents [508].

Studies show that maternal behavior is promoted by central oxytocin administration, and antagonists of the peptide action can reverse these effects [509]. However, central administration of oxytocin can also induce the labor and lactation effects elicited by oxytocin release from the posterior pituitary [510, 511]. These results suggest that the roles of oxytocin in the brain and pituitary overlap. In addition, female mice that lack oxytocin exhibit normal parturition but do not eject milk [512] which suggests an interesting synergy between the central and peripheral effects of the neuropeptide, especially with regard to birth and the subsequent rearing of offspring [512]. Interestingly, lack of oxytocin has been recently linked to several psychiatric phenomena, including sociopathy, psychopathy, and narcissism [513] (Table 3.?). Thus oxytocin is a neuropeptide for which many functions are still being characterized, and may yet reveal more physiological roles.

Oxytocin receptors

Oxytocin exerts its physiological effects by activating the oxytocin receptor although it can also bind to the closely related vasopressin receptor subtypes [514]. The oxytocin receptor gene (*OXTR*) has several alleles that give rise to receptors of differing effectiveness. The "G" allele imparts higher empathy, as well as lower response to stress [515]. Oxytocin receptors are expressed in many areas of the brain including the amygdala, the ventromedial hypothalamus, the septum, the nucleus accumbens, and the brainstem. These receptors are also found in the spinal cord [516].

Oxytocin receptors belong to the family A of GPCRs and couple to Gαq proteins; thus the binding of a receptor agonist leads to activation of PLC to produce the second messengers DAG and IP3 [506]. The signaling pathways mediated by Gαq-coupled receptors are described in Chapter 3.

5.2.6 VASOPRESSIN & VASOPRESSIN RECEPTORS

Vasopressin

Vasopressin (also known as arginine vasopressin and antidiuretic hormone) is a nonapeptide that has a structure very similar to that of oxytocin differing from the latter by two amino acid residues. In fact, the genes for both peptides are located in the same chromosome and separated by a relatively short distance. Vasopressin is generated by the proteolytic cleavage of a precursor that also yields neurophysin II and a 32 amino acid glycopeptide [504]. The vasopressin precursor is synthesized in the hypothalamus and is stored in vesicles at the posterior pituitary to be released into the bloodstream. The primary characterized function of vasopressin in the periphery is the regulation of water absorption in the collecting duct of the kidney nephron. Vasopressin mediates the permeability of the membranes of these cells by promoting the translocation of aquaporin channels to the plasma membrane [517]. Thus vasopressin plays a crucial role for water retention during dehydration, hence it's other name—antidiuretic hormone. In addition, vasopressin is intimately involved in the regulation of the levels of salts, glucose, and other blood components, and it regulates blood pressure by moderating vasoconstriction, the role from which it derives its name. Vasopressin released directly into the brain is important for mediation of pair bonding behavior in a gender specific manner [508]. Vasopressin released from the neurons of the supraoptic nucleus of the hypothalamus may play a role not only in pair bonding behavior but also in the regulation of body temperature, aggression and analgesia; surprisingly the analgesic effects of vasopressin are gender specific [518, 519]. Another interesting aspect of vasopressin function is how it gives rise to different behaviors in males depending upon the gender of the counterpart to which they are presented. For example, vasopressin levels increase in the male brain during sexual interaction with a female to promote pair bond formation, while interaction with another male promotes aggressive behavior [508, 519]. The involvement of vasopressin in social behaviors is supported by studies showing that central administration of vasopressin facilitates changes in male parental behavior in the monogamous prairie vole, inhibiting aggression directed toward the pups, even in non-paternal males [520]. Conversely, vasopressin receptor antagonists blocked paternal behaviors and promoted aggressive behaviors toward the pups [520].

Vasopressin receptors

Vasopressin exerts its biological effects through activation of vasopressin receptors although it can also bind to the closely related oxytocin receptor [514]. Three subtypes of the vasopressin receptor have been identified: vasopressin V1A, V1B and V2 that are encoded by three different genes *AVPR1A*, *AVPR1B*, and *AVPR2* [521]. These receptors exhibit unique tissue distribution with vasopressin V1A receptor being expressed in vascular smooth muscle cells, hepatocytes, platelets, brain and uterine cells, vasopressin V1B receptor expressed throughout the brain particularly in the cells of the anterior pituitary and hippocampal pyramidal cells while vasopressin V2 receptor expressed in kidneys, lungs and liver [521].

Vasopressin receptors belong to the family A of GPCRs. Vasopressin receptors exhibit differences in coupling to G proteins. For example, vasopressin V1A and V1B receptors are coupled to Gαq protein and their activation leads to the stimulation of PLC activity to produce IP3 and DAG [522]. In contrast, vasopressin V2 receptors are coupled to Gαs and its activation leads to increases in cyclic AMP levels and activation of down-stream signaling cascades (see Chapter 3). Vasopressin V2 receptors located at the basolateral surface of epithelial cells of the nephron play a role in water retention by promoting a cyclic AMP-mediated insertion of aquaporin-2 channels into the plasma membrane [522].

Vasopressin receptors also play an important role in mediating many important physiological processes, both centrally and peripherally. The brain localization of the vasopressin receptors correlates with their proposed function is regulating memory, aggression, and other behaviors while its localization to kidneys correlates with its function in modulating water retention [523]. The major peripheral role of vasopressin (or antidiuretic hormone) is to maintain water balance by affecting water resorption from the kidneys. Studies with mice lacking vasopressin receptors support their involvement in these physiological processes. For example, mice lacking the vasopressin V1A receptor exhibit reduced anxiety-like behaviors and greatly impaired social recognition skills without any defects in spatial and nonsocial olfactory learning and memory [519, 524]. Some of the mice also exhibit deficits in their circadian rhythms and olfaction [519]. Mice lacking the vasopressin V1B receptor exhibit reduced levels of social aggressive behavior and mildly impaired social recognition [519].

5.2.7 PEPTIDE ENDOCANNABINOIDS AND CANNABINOID RECEPTORS

Peptide endocannabinoids

The most well-characterized endocannabinoids are small lipophilic molecules derived from arachidonic acid, anandamide and 2-arachidonylglycerol (2-AG). These lipid-derived endocannabinoids are involved in many physiological processes including memory formation, appetite, stress response, and immune response modulation [525, 526]. Recently peptide modulators of cannabinoid receptors known as hemopressins (Hp) have been identified.

Hemopressin peptides are generated by non-classical processing of the α chain of hemoglobin by hitherto unknown mechanisms. The peptide with the sequence PVNFKFLSH, named hemopressin, was originally identified in extracts of rat brain and shown to function as an inverse agonist of CB1cannabinoid receptors while longer forms of this peptide (the N terminally extended RVD-hemopressin and VD-hemopressin) acted as receptor agonists [527]. Hemopressin peptides derived from the β chain of hemoglobin that exhibit agonistic activity at CB1 cannabinoid receptors have also been identified [527]. These peptides with agonistic activity are consistently identified in mass spectrometric analysis of different brain regions [347, 527, 528]. Recent studies reporting the presence of α- and β-chains of hemoglobin in neurons suggest that these peptides are generated in the brain and secreted from neurons by an as yet uncharacterized mechanism [529, 530]. Interestingly, the shorter hemopressin peptide appears to modulate the CB1 receptor-mediated effects on analgesia and feeding by antagonizing receptor function [347, 531–533].

Cannabinoid receptors

Endocannabinoids modulate many of their physiological effects through the activation of cannabinoid receptors. Two well-described subtypes of the cannabinoid receptor are the CB1 and

CB2 receptors that are encoded by the *CNR1* and *CNR2* genes respectively [534]. The CB1 and CB2 cannabinoid receptors exhibit differential tissue distribution. CB1 cannabinoid receptors are expressed mainly in the brain and central nervous system, but also in the lungs, liver, kidneys, and adipose tissue. The CB1 receptors in the brain are responsible for the psychotropic effects of THC, the active ingredient of marijuana [535]. CB2 cannabinoid receptors are mainly found in the immune system and in hematopoietic cells more specifically in immune T cells, macrophages and B cells. There is also evidence that these receptors are expressed on peripheral nerves, where they may be involved in pain perception [536].

Both CB1 and CB2 cannabinoid receptors belong to the family A of GPCRs and activate Gαi-mediated signaling cascades leading to the inhibition of adenylyl cyclase activity and a decrease in cyclic AMP levels. Activation of CB1 cannabinoid receptors by longer hemopressin peptides leads to the activation of signaling pathways distinct from the pathway activated by classic cannabinoid ligands [527].

Mice lacking cannabinoid receptors show that CB1 receptors are involved in the regulation of several behavioral responses including locomotion, anxiety- and depressive-like states, cognitive functions such as memory and learning processes, cardiovascular responses, pain attenuation and feeding [537]. Interestingly, these animals maintain sensitivity to THC, indicating that this phytocannabinoid exerts its function both through the CB1 and CB2 receptors [538]. Studies with mice lacking CB2 cannabinoid receptors indicate that this receptor plays a role in immune cell function and in embryonic development, bone loss, liver disorders, peripheral pain, autoimmune inflammation, atherosclerosis, apoptosis and chemotaxis [539].

5.3 ENZYME-LINKED RECEPTORS

Enzyme-linked receptors (Trk receptors and receptors for a variety of growth factors) are single transmembrane proteins with an extracellular ligand binding domain, a transmembrane domain, and a cytosolic domain, that in many cases, functions as a tyrosine kinase, hence the name *receptor tyrosine kinases* [406]. In some cases, the cytosolic domain of the receptor protein is devoid of enzyme activity but is able to associate with intracellular enzymes [406]. For both of these receptor groups, the binding of the neuropeptide ligand to the extracellular domain leads to a conformational change resulting in the formation/stabilization of receptor dimers [406]. This leads to the interaction of intracellular domains causing the phosphorylation of tyrosine residues in the cytosolic domain [406] (Fig. 5.1). Once phosphorylated, these domains recruit other signaling molecules, including kinases that phosphorylate other cellular substrates and produce an intracellular response [406]. In this section we briefly describe this receptor system with a few specific examples.

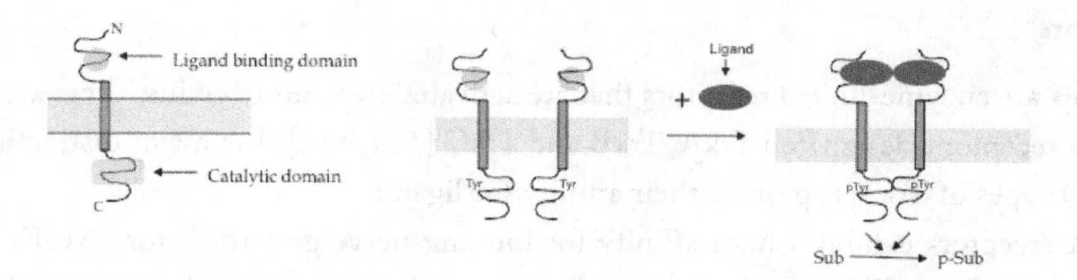

Figure 5.1. **Structure and activation mechanism of enzyme-linked receptors.** The rectangular segment in each receptor subunit represents the membrane spanning α-helical region made up of approximately 20 amino acids. Binding of the ligand leads to phosphorylation of the catalytic domain; the latter recruits and phosphorylates downstream substrate molecules leading to an intracellular response. Tyr, tyrosine, p-Tyr, phosphorylated tyrosine, Sub, substrate; p-Sub, phosphorylated substrate.

Neurotrophins

Neurotrophins is a term used to designate four structurally related peptides, nerve growth factor (NGF), neurotrophin-3 (NT-3), neurotrophin-4 (NT-4) and BDNF. These peptides are processed from precursor proteins, Pro-NGF, pro-NT-3, pro-NT-4 and pro-BDNF, either inside the cell by furin or PCs, or extracellularly by plasmin or the matrix metalloproteinases [540]. Pro-NGF, pro-NT-3, and pro-NT-4 are packaged into constitutive vesicles before secretion whereas pro-BDNF is primarily packaged into regulated secretory pathway vesicles, processed and secreted in an activity-dependent manner [540–542].

NGF is necessary for the survival and maintenance of sympathetic and sensory neurons since without it, these neurons undergo apoptosis [543]. NGF also plays a role in axonal growth, branching and elongation [544]. NT-3 plays an important role in the development and maintenance of the nervous system, in the regulation of monoamine neurotransmitters such as serotonin and noradrenaline, and in enhancing NGF and BDNF levels [545]. NT-4 is involved with the development of long-term memory [546], and has also been implicated in the regulation of appetite and body weight [547]. BDNF helps to support the survival of existing neurons, and promotes the growth and differentiation of new neurons and synapses [548, 549]; it also plays an important role in long-term memory [550].

Neurotrophins bind and activate distinct tyrosine kinase receptors (Trks): NGF binds TrkA, BDNF and NT-4 bind TrkB, and NT-3 binds TrkC [540]. In addition, neurotrophins bind and activate a pan-neurotrophin receptor, p75NTR [540]. The Trk receptors are described below.

Trk receptors

Trk receptors are enzyme-linked receptors that are activated by neurotrophins. There are three main types of Trk receptors, designated TrkA, TrkB and TrkC [551, 552]. The major distinction between the three subtypes of Trk receptors is their affinity for ligands.

TrkA receptors exhibit a high affinity for binding nerve growth factor (NGF) [553, 554]. The binding of NGF by TrkA has been shown to play a role in the expression and regulation of enzymes necessary for the generation of neurotransmitters, in addition to regulating the motility and direction of neuritic growth cones as well as a variety of nuclear functions [555]. TrkA receptors are predominantly expressed in the relatively thinner axons of nociceptive sensory neurons [556–558].

TrkB receptors exhibit the highest affinity for the binding of NT-4 and BDNF [554, 559]. TrkB receptors are most active in those areas of the central nervous system that subserve the processes associated with learning, memory, and higher-order cognitive capabilities [560, 561]. The primary role performed by BDNF in the central nervous systems is to support the survival of neurons [562–565]. Given its well-documented role in nurturing new synapses, TrkB/BDNF is thought to play a critical role in the formation of long-term memory in the central nervous system [566–568]. TrkB/BDNF are also known to play a critical role in neural development [569].

TrkC receptors are activated by NT-3, and show little if any response to any other ligands [570]. TrkC is predominantly expressed in the very thick axons of proprioceptive sensory neurons in the peripheral nervous system [556, 571].

5.4 PERIPHERALLY SYNTHESIZED PROTEINS/PEPTIDES AND THEIR RECEPTORS IN THE BRAIN

As already mentioned in section 5.1 neuronal function in the brain can also be modulated by peptides generated in the periphery which bind to their receptors present in the brain. Examples for such peptides include leptin and insulin that bind to enzyme-linked receptors. These are described below.

5.4.1 LEPTIN AND THE LEPTIN RECEPTOR

Leptin

Leptin is a 167 amino acid protein that is encoded by the *Lep* gene that was previously known as the *ob* gene [572–574]. Leptin is primarily synthesized by white adipose tissue [572, 573], brown adipose tissue, stomach, liver, placenta, ovaries, skeletal muscle, mammary epithelial cells, bone marrow, pituitary gland, and brain [573, 575, 576]. Leptin plays a role in a number of important physiological processes such as central appetite inhibition, suppression of lipogenesis, modulation of the activity of T-cells and of the mammalian ovulatory cycle, promotion of angiogenesis, regulation of bone metabolism and inflammatory responses as well as modulating stress responses via its effects on the hypothalamo-pituitary adrenal axis [573, 577, 578].

Following release into the circulation, leptin crosses the blood-brain-barrier via an as yet uncharacterized saturable mechanism involving the short receptor form, LepRe, and binds to presynaptic GABAergic neurons [579–581]. Leptin modulates energy balance via its effects on NPY/AgRP and POMC/CART containing neurons in the arcuate nucleus of the hypothalamus [582]. Increased levels of leptin, that are indicative of a fed state with sufficient energy stores, will cause an increase in POMC/CART expression and reduce NPY/AgRP levels in the arcuate nucleus [582]. Concomitantly, decreased levels of leptin are indicative of a negative energy balance and will cause a decrease in POMC/CART expression, an increase in NPY/AgRP levels leading to an increase in feeding [582]. Interestingly, lack of leptin in humans leads to a constant desire for food and ultimately severe obesity [583, 584]. Leptin exerts its effects by binding to receptors that resemble the class I cytokine receptor family. These are described below.

Leptin receptors

The leptin receptor (LepR) is a single transmembrane non-enzyme linked receptor with extracellular, TM, and cytoplasmic domains [582]. Based on the length of the intracellular domain the different leptin receptor isoforms are characterized as short isoforms (LepRa, c, d, and f) where the cytoplasmic domain ranges from 32–40 amino acids and a long isoform (LepRb) with a cytoplasmic domain comprising 302 amino acids. The isoform that lacks the TM domain (LepRe) is considered to be the secreted form of the leptin receptor that helps transport leptin across the blood-brain barrier.

Binding of leptin to the leptin receptor, LepRb, leads to receptor dimerization followed by the docking of Janus kinase (JAK) and activation of JAK2 by autophosphorylation at tyrosine residues [573, 582]. Activated JAK2 phosphorylates LepRb leading to the recruitment of STAT proteins, particularly STAT3, to the receptor. Recruited STAT3 proteins are phosphorylated leading to their dimerization following which they are translocated to the nucleus where they modulate the transcription of genes that mediate the effects of leptin [573, 582]. In addition to the JAK/STAT3 pathway, LepRb can signal via the ERK1/2 and PI3K pathways [582]. Moreover, LepRb-mediated signaling is under negative feedback control by suppressors of cytokine signaling proteins (SOCS) particularly SOCS3 [582]. Leptin receptors can be detected in the pituitary and brain, particularly in the brainstem, hypothalamus, as well as in heart, liver, lungs, pancreas, ovaries, testis, skeletal muscle and hemopoietic organs [573, 582].

Mice with mutations in the *LepR* gene are hyperphagic and severely obese supporting the involvement of leptin in regulation of feeding and energy homeostasis. These mice also exhibit a number of metabolic and endocrine abnormalities including diabetes, hypercortisolemia, infertility, cold intolerance and enlarged steatotic livers [585].

5.4.2 INSULIN AND THE INSULIN RECEPTOR

Insulin

Insulin is a peptide hormone that is synthesized in mammals in the β-cells of the islets of Langerhans in the pancreas from proinsulin [586, 587]. Processing of insulin leads to the generation of A-chain and B-chain, that are linked by disulfide bonds [586]. PCs and CPE are involved in the processing of proinsulin into insulin [588]. Mature insulin is packaged into mature secretory vesicles waiting for metabolic signals, such as glucose and vagal nerve stimulation, to be secreted from the cell into the circulation [589]. Insulin plays a central role in the regulation of carbohydrate and fat metabolism in the body. Moreover, insulin has profound effects in the central nervous system, where it regulates key processes such as energy homeostasis, reproductive endocrinology and neuronal survival.

Increases in blood glucose levels lead to the secretion of insulin which, in turn, causes cells in liver, skeletal muscle and fat tissue to take up glucose from blood and store it as glycogen (liver, skeletal muscle) or triglycerides (fat tissue) [589]. In addition to glucose, insulin secretion can be increased by the amino acids arginine and leucine, the parasympathetic release of acetylcholine, cholecystokinin, glucagon-like-peptide-1 and glucose-dependent insulinotropic peptide [589, 590]. There are two stages of insulin secretion by the β-cells of the islets of Langerhans:(i) a rapid phase that is triggered in response to increased glucose levels and is the primary method of insulin secretion; and (ii) a slow sustained secretion of insulin from newly formed vesicles that is independent of glucose levels [589]. The rapid phase of insulin secretion has been shown to be impaired in patients with type II diabetes [591].

The primary sites for clearance of insulin from the circulation are the liver and kidneys. Degradation of insulin involves endocytosis of the insulin-receptor complex, followed by the action of insulin-degrading enzyme [592]. Disturbances in insulin levels lead to pathological conditions. For example, lack of insulin due to destruction of β-cells of the islets of Langerhans leads to type I diabetes while impaired insulin activity due to development of insulin resistance can lead to type II diabetes [591]. Insulin exerts its biological effects by binding to cell surface receptors that belong to the class of enzyme-linked receptors. These are described below.

Insulin receptors

The insulin receptor, a tyrosine kinase receptor, is synthesized as a pro-receptor that undergoes processing and dimerization in the early secretory compartment [593]. In the Golgi the dimerized polypeptide chains are cleaved into α- and β-subunits that are linked via disulfide bonds to form a tetrameric receptor comprising of two identical extracellular α-subunits that bind insulin and two identical TM β-subunits that have cytoplasmic tyrosine kinase activity [593].

The binding of insulin to the extracellular α-subunits leads to a conformational change that induces the autophosphorylation of specific tyrosine residues on the β-subunits; this, in turn, lead to further conformational changes that activate the receptor's tyrosine kinase activity [593]. The latter then leads to the phosphorylation of down-stream insulin receptor substrates that serve as docking molecules that bind to and activate cellular kinases that mediate the cellular effects of insulin [593]. The effects of insulin on glucose transport and on the metabolic effects of insulin are mediated via the PI3K-mediated phosphorylation of IRS-1 while effects on cell growth and protein synthesis involve the Ras/MAPK pathway [593]. Insulin receptors are ubiquitously distributed, but within the brain highest levels are observed in the olfactory bulb, cerebral cortex, hypothalamus, cerebellum and choroid plexus [594].

Studies suggest that insulin can be transported across the blood-brain-barrier although very little is known about the mechanism involved [595]. A number of factors can affect the transport of insulin through the blood-brain-barrier; these include fasting, obesity, aging or diabetes mellitus [595]. In the brain the binding of insulin to receptors present in the arcuate nucleus of the hypothalamus not only leads to a decrease in food intake, via modulation of the expression levels of hypothalamic peptides involved in modulation of feeding behaviors, but also to increases in insulin sensitivity in peripheral tissues [595]. Interestingly, mice with a targeted disruption of the insulin receptor in the brain exhibit mild diet-sensitive obesity and reduced fertility [596].

CHAPTER 6

Perspectives

This book describes GPCRs, the structural information obtained by recent crystallographic studies of the antagonist bound receptor, the signaling pathways activated by these receptors as well as gives specific examples of a few neuropeptide receptor systems. The function of GPCRs can be regulated by interaction with other proteins or ligands in several ways, including, and not limited to the following:

1. Interaction with molecular chaperones to induce or reduce receptor trafficking to the cell surface under certain conditions;
2. Interaction with other receptors at the plasma membrane to cause alterations in the binding and signaling properties of both receptors;
3. Interaction with selective agonists to stabilize unique receptor conformations and induce selective signaling; and
4. Interaction with selective adaptor/sorting proteins to direct receptors to degradative or cell surface recycling pathways.

Information about the molecular mechanism of these interactions may allow for the development of better drugs targeting GPCRs involved in pathological conditions such as those that enhance the function of analgesia-promoting GPCRs in the setting of neuropathic pain.

In recent years it has been discovered that GPCRs can respond differently to different agonists. This phenomenon, termed "biased agonism" or "ligand directed signaling," is associated with the promotion of distinct signaling or receptor trafficking pathways by the same receptor in response to one agonist compared to a different agonist. In this context compounds that stabilize distinct receptor conformations, thereby eliciting the activation of select signal transduction pathways, could serve as a new generation of therapeutics with reduced side effects.

Another potential strategy for the development of a new generation of therapeutics involves targeting receptor heterodimers. It is now clear that receptor heterodimers can play a key role in disease processes. Targeting disease-specific heterodimers is likely to have several benefits:

1. Fewer off-target effects, as the heterodimer is only upregulated in regions whose receptor activity is pathologically altered;

2. An increase in drug potency, compared to ligands for individual receptors (e.g., heterodimer-specific compounds should achieve the desired effect at the heterodimer at lower doses than at homodimers); and

3. A potentiation or reduction in heteromer-specific signaling, if it is protective or detrimental, respectively, during a disease state.

The current challenge is to find or design compounds that are heteromer specific. The design of bivalent ligands—that is, ligands composed of two pharmacophores spaced by a linker region and targeting two receptors within a complex—could be useful in this setting. Another strategy is to use heteromer-selective antibodies as drugs. Such antibodies have been recently generated for μ-δ opioid [597], δ-κ opioid [598], and CB1R-AT1R [599] heterodimers.

Neuropeptide receptors are involved in a number of disorders (e.g., pain, obesity, addiction, depression) and given that GPCRs represent ~30% of drug targets, a considerable amount of time and effort has been spent on identifying drugs targetting these receptors. Despite these efforts few drugs have been identified. This has led to a renewed interest in identifying new GPCRs or identifying new pharmacologies emerging from known GPCRs (such as biased agonism, dimerization or receptor-interacting proteins). These are likely to serve as new targets for the identification of new therapeutics (drugs and biologicals) toward the treatment of disorders involving neuropeptide receptors.

References

1. De Carlos JA, Borrell J (2007) A historical reflection of the contributions of Cajal and Golgi to the foundations of neuroscience. *Brain Res Rev* 55: 8-16. DOI: 10.1016/j.brainresrev.2007.03.010

2. Raju TN (1999) The Nobel chronicles. 1936: Henry Hallett Dale (1875-1968) and Otto Loewi (1873-1961). *Lancet* 353: 416. DOI: 10.1016/S0140-6736(05)75001-7

3. Maehle AH (2004) "Receptive substances": John Newport Langley (1852-1925) and his path to a receptor theory of drug action. *Med Hist* 48: 153-174. DOI: 10.1017/S0025727300000090

4. Maehle AH, Prull CR, Halliwell RF (2002) The emergence of the drug receptor theory. *Nat Rev Drug Discov* 1: 637-641. DOI: 10.1038/nrd875

5. Kosterlitz HW (1980) Opioid peptides and their receptors. *Prog Biochem Pharmacol* 16: 3-10.

6. Hughes J, Smith TW, Kosterlitz HW, Fothergill LA, Morgan BA, et al. (1975) Identification of two related pentapeptides from the brain with potent opiate agonist activity. *Nature* 258: 577-580. DOI: 10.1038/258577a0

7. Kosterlitz HW (1987) Endogenous opioids and their receptors. *Pol J Pharmacol Pharm* 39: 571-576.

8. Wess J (1997) G-protein-coupled receptors: molecular mechanisms involved in receptor activation and selectivity of G-protein recognition. *FASEB J* 11: 346-354.

9. Hedin KE, Duerson K, Clapham DE (1993) Specificity of receptor-G protein interactions: searching for the structure behind the signal. *Cell Signal* 5: 505-518. DOI: 10.1016/0898-6568(93)90046-O

10. Deupi X, Standfuss J (2011) Structural insights into agonist-induced activation of G-protein-coupled receptors. *Curr Opin Struct Biol* 21: 541-551. DOI: 10.1016/j.sbi.2011.06.002

11. Fredriksson R, Lagerstrom MC, Lundin LG, Schioth HB (2003) The G-protein-coupled receptors in the human genome form five main families. Phylogenetic analysis, paralogon groups, and fingerprints. *Mol Pharmacol* 63: 1256-1272. DOI: 10.1124/mol.63.6.1256

12. IHGS C (2001) Initial sequencing and analysis of the human genome. *Nature* 409: 860-922. DOI: 10.1038/35057062

13. Klabunde T, Hessler G (2002) Drug design strategies for targeting G-protein-coupled receptors. *Chembiochem* 3: 928-944. DOI: 10.1002/1439-7633(20021004)3:10<928::AID-CBIC928>3.0.CO;2-5

14. Seifert R, Wenzel-Seifert K (2002) Constitutive activity of G-protein-coupled receptors: cause of disease and common property of wild-type receptors. *N-S Arch Pharmacol* 366: 381-416. DOI: 10.1007/s00210-002-0588-0

15. Kobilka BK (2007) G protein coupled receptor structure and activation. *Biochim Biophys Acta* 1768: 794-807. DOI: 10.1016/j.bbamem.2006.10.021

16. Gether U (2000) Uncovering molecular mechanisms involved in activation of G protein-coupled receptors. *Endocr Rev* 21: 90-113. DOI: 10.1210/er.21.1.90

17. Lu ZL, Saldanha JW, Hulme EC (2002) Seven-transmembrane receptors: crystals clarify. *Trends Pharmacol Sci* 23: 140-146. DOI: 10.1016/S0165-6147(00)01973-8

18. Marinissen MJ, Gutkind JS (2001) G-protein-coupled receptors and signaling networks: emerging paradigms. *Trends Pharmacol Sci* 22: 368-376. DOI: 10.1016/S0165-6147(00)01678-3

19. Malbon CC, Wang H, Moon RT (2001) Wnt signaling and heterotrimeric G-proteins: strange bedfellows or a classic romance? *Biochem Biophys Res Commun* 287: 589-593. DOI: 10.1006/bbrc.2001.5630

20. Attwood TK, Findlay JB (1994) Fingerprinting G-protein-coupled receptors. *Protein Eng* 7: 195-203. DOI: 10.1093/protein/7.2.195

21. Ji TH, Grossmann M, Ji I (1998) G protein-coupled receptors. I. Diversity of receptor-ligand interactions. *J Biol Chem* 273: 17299-17302. DOI: 10.1074/jbc.273.28.17299

22. DeGraff JL, Gurevich VV, Benovic JL (2002) The third intracellular loop of alpha 2-adrenergic receptors determines subtype specificity of arrestin interaction. *J Biol Chem* 277: 43247-43252. DOI: 10.1074/jbc.M207495200

23. Okamoto Y, Ninomiya H, Tanioka M, Sakamoto A, Miwa S, et al. (1997) Palmitoylation of human endothelinB. Its critical role in G protein coupling and a differential requirement for the cytoplasmic tail by G protein subtypes. *J Biol Chem* 272: 21589-21596. DOI: 10.1074/jbc.272.34.21589

24. Palmer TM, Stiles GL (2000) Identification of threonine residues controlling the agonist-dependent phosphorylation and desensitization of the rat A(3) adenosine receptor. *Mol Pharmacol* 57: 539-545.

25. Katritch V, Cherezov V, Stevens RC (2012) Diversity and modularity of G protein-coupled receptor structures. *Trends Pharmacol* Sci 33: 17-27. DOI: 10.1016/j.tips.2011.09.003

26. Escriba PV, Wedegaertner PB, Goni FM, Vogler O (2007) Lipid-protein interactions in GPCR-associated signaling. *Biochim Biophys Acta* 1768: 836-852. DOI: 10.1016/j.bbamem.2006.09.001

27. van Rhee AM, Jacobson KA (1996) Molecular architecture of G protein-coupled receptors. *Drug Dev Res* 37: 1-38. DOI: 10.1002/(SICI)1098-2299(199601)37:1<1::AID-DDR1>3.0.CO;2-S

28. Rovati GE, Capra V, Neubig RR (2007) The highly conserved DRY motif of class A G protein-coupled receptors: beyond the ground state. *Mol Pharmacol* 71: 959-964. DOI: 10.1124/mol.106.029470

29. Tiburu EK, Tyukhtenko S, Zhou H, Janero DR, Struppe J, et al. (2011) Human cannabinoid 1 GPCR C-terminal domain interacts with bilayer phospholipids to modulate the structure of its membrane environment. *AAPS J* 13: 92-98. DOI: 10.1208/s12248-010-9244-7

30. Schoneberg T, editor (2002) *GPCR Superfamily and its structural characterization*. New York: Oxford University Press. 3-27 p.

31. Devi LA (2005) *The G protein-coupled receptors handbook*: Humana Press, Totowa, N.J. DOI: 10.1007/978-1-59259-919-6

32. George SR, O'Dowd BF, Lee SP (2002) G-protein-coupled receptor oligomerization and its potential for drug discovery. *Nat Rev Drug Discov* 1: 808-820. DOI: 10.1038/nrd913

33. Strader CD, Fong TM, Graziano MP, Tota MR (1995) The family of G-protein-coupled receptors. *FASEB J* 9: 745-754.

34. Harmar AJ (2001) Family-B G-protein-coupled receptors. *Genome Biol 2*: reviews 3013.1-reviews 3013.10.

35. Ulrich CD, 2nd, Holtmann M, Miller LJ (1998) Secretin and vasoactive intestinal peptide receptors: members of a unique family of G protein-coupled receptors. *Gastroenterology* 114: 382-397. DOI: 10.1016/S0016-5085(98)70491-3

36. Kunishima N, Shimada Y, Tsuji Y, Sato T, Yamamoto M, et al. (2000) Structural basis of glutamate recognition by a dimeric metabotropic glutamate receptor. *Nature* 407: 971-977. DOI: 10.1038/35039564

37. Pin JP, Comps-Agrar L, Maurel D, Monnier C, Rives ML, et al. (2009) G-protein-coupled receptor oligomers: two or more for what? Lessons from mGlu and GABAB receptors. *J Physiol* 587: 5337-5344. DOI: 10.1113/jphysiol.2009.179978

38. Ray K, Hauschild BC (2000) Cys-140 is critical for metabotropic glutamate receptor-1 dimerization. *J Biol Chem* 275: 34245-34251. DOI: 10.1074/jbc.M005581200

39. Ray K, Hauschild BC, Steinbach PJ, Goldsmith PK, Hauache O, et al. (1999) Identification of the cysteine residues in the amino-terminal extracellular domain of the human Ca(2+) receptor critical for dimerization. Implications for function of monomeric Ca(2+) receptor. *J Biol Chem* 274: 27642-27650. DOI: 10.1074/jbc.274.39.27642

40. Margeta-Mitrovic M, Jan YN, Jan LY (2000) A trafficking checkpoint controls GABA(B) receptor heterodimerization. *Neuron* 27: 97-106. DOI: 10.1016/S0896-6273(00)00012-X

41. Urwyler S (2011) Allosteric modulation of family C G-protein-coupled receptors: from molecular insights to therapeutic perspectives. *Pharmacol Rev* 63: 59-126. DOI: 10.1124/pr.109.002501

42. Tsuchiya D, Kunishima N, Kamiya N, Jingami H, Morikawa K (2002) Structural views of the ligand-binding cores of a metabotropic glutamate receptor complexed with an antagonist and both glutamate and Gd3+. *Proc Natl Acad Sci USA* 99: 2660-2665. DOI: 10.1073/pnas.052708599

43. Kniazeff J, Galvez T, Labesse G, Pin JP (2002) No ligand binding in the GB2 subunit of the GABA(B) receptor is required for activation and allosteric interaction between the subunits. *J Neurosci* 22: 7352-7361.

44. Granier S, Manglik A, Kruse AC, Kobilka TS, Thian FS, et al. (2012) Structure of the delta-opioid receptor bound to naltrindole. *Nature* 485: 400-404. DOI: 10.1038/nature11111

45. Manglik A, Kruse AC, Kobilka TS, Thian FS, Mathiesen JM, et al. (2012) Crystal structure of the micro-opioid receptor bound to a morphinan antagonist. *Nature* 485: 321-326. DOI: 10.1038/nature10954

46. Wu B, Chien EY, Mol CD, Fenalti G, Liu W, et al. (2010) Structures of the CXCR4 chemokine GPCR with small-molecule and cyclic peptide antagonists. *Science* 330: 1066-1071. DOI: 10.1126/science.1194396

47. Wu H, Wacker D, Mileni M, Katritch V, Han GW, et al. (2012) Structure of the human kappa-opioid receptor in complex with JDTic. *Nature* 485:327-332. DOI: 10.1038/nature10939

48. Hanson MA, Stevens RC (2009) Discovery of new GPCR biology: one receptor structure at a time. *Structure* 17: 8-14. DOI: 10.1016/j.str.2008.12.003

49. Zhou H, Tai HH (2000) Expression and functional characterization of mutant human CXCR4 in insect cells: role of cysteinyl and negatively charged residues in ligand binding. *Arch Biochem Biophys* 373: 211-217. DOI: 10.1006/abbi.1999.1555

50. George SR, Fan T, Xie Z, Tse R, Tam V, et al. (2000) Oligomerization of mu- and delta-opioid receptors. Generation of novel functional properties. *J Biol Chem* 275: 26128-26135. DOI: 10.1074/jbc.M000345200

51. Shimohigashi Y, Waki M, Izumiya N, Costa T, Herz A, et al. (1986) Opiate receptor binding characteristics of dimeric analogues of mu-selective DAGO-enkephalin. *Biochem Int* 13: 199-203.

52. Wang D, Sun X, Bohn LM, Sadee W (2005) Opioid receptor homo- and heterodimerization in living cells by quantitative bioluminescence resonance energy transfer. *Mol Pharmacol* 67: 2173-2184. DOI: 10.1124/mol.104.010272

53. Filizola M, Devi LA (2012) Grand opening of structure-guided design for novel opioids. *Trends Pharmacol Sci* 34: 6-12. DOI: 10.1016/j.tips.2012.10.002

54. Thompson AA, Liu W, Chun E, Katritch V, Wu H, et al. (2012) Structure of the nociceptin/orphanin FQ receptor in complex with a peptide mimetic. *Nature* 485: 395-399. DOI: 10.1038/nature11085

55. Goldsmith PK, Fan GF, Ray K, Shiloach J, McPhie P, et al. (1999) Expression, purification, and biochemical characterization of the amino-terminal extracellular domain of the human calcium receptor. *J Biol Chem* 274: 11303-11309. DOI: 10.1074/jbc.274.16.11303

56. Pace AJ, Gama L, Breitwieser GE (1999) Dimerization of the calcium-sensing receptor occurs within the extracellular domain and is eliminated by Cys --> Ser mutations at Cys101 and Cys236. *J Biol Chem* 274: 11629-11634. DOI: 10.1074/jbc.274.17.11629

57. AbdAlla S, Zaki E, Lother H, Quitterer U (1999) Involvement of the amino terminus of the B(2) receptor in agonist-induced receptor dimerization. *J Biol Chem* 274: 26079-26084. DOI: 10.1074/jbc.274.37.26079

58. Abe J, Suzuki H, Notoya M, Yamamoto T, Hirose S (1999) Ig-hepta, a novel member of the G protein-coupled hepta-helical receptor (GPCR) family that has immunoglobulin-like repeats in a long N-terminal extracellular domain and defines a new subfamily of GPCRs. *J Biol Chem* 274: 19957-19964. DOI: 10.1074/jbc.274.28.19957

59. Kuner R, Kohr G, Grunewald S, Eisenhardt G, Bach A, et al. (1999) Role of heteromer formation in GABAB receptor function. *Science* 283: 74-77. DOI: 10.1126/science.283.5398.74

60. White JH, Wise A, Main MJ, Green A, Fraser NJ, et al. (1998) Heterodimerization is required for the formation of a functional GABA(B) receptor. *Nature* 396: 679-682. DOI: 10.1038/25354

61. Cvejic S, Devi LA (1997) Dimerization of the delta opioid receptor: implication for a role in receptor internalization. *J Biol Chem* 272: 26959-26964. DOI: 10.1074/jbc.272.43.26959

62. Fotiadis D, Liang Y, Filipek S, Saperstein DA, Engel A, et al. (2003) Atomic-force microscopy: Rhodopsin dimers in native disc membranes. *Nature* 421: 127-128. DOI: 10.1038/421127a

63. George SR, Lee SP, Varghese G, Zeman PR, Seeman P, et al. (1998) A transmembrane domain-derived peptide inhibits D1 dopamine receptor function without affecting receptor oligomerization. *J Biol Chem* 273: 30244-30248. DOI: 10.1074/jbc.273.46.30244

64. Hebert TE, Bouvier M (1998) Structural and functional aspects of G protein-coupled receptor oligomerization. *Biochem Cell Biol* 76: 1-11. DOI: 10.1139/o98-012

65. Liang Y, Fotiadis D, Filipek S, Saperstein DA, Palczewski K, et al. (2003) Organization of the G protein-coupled receptors rhodopsin and opsin in native membranes. *J Biol Chem* 278: 21655-21662. DOI: 10.1074/jbc.M302536200

66. Hebert TE, Moffett S, Morello JP, Loisel TP, Bichet DG, et al. (1996) A peptide derived from a beta2-adrenergic receptor transmembrane domain inhibits both receptor dimerization and activation. *J Biol Chem* 271: 16384-16392. DOI: 10.1074/jbc.271.27.16384

67. Ng GY, O'Dowd BF, Lee SP, Chung HT, Brann MR, et al. (1996) Dopamine D2 receptor dimers and receptor-blocking peptides. *Biochem Biophys Res Commun* 227: 200-204. DOI: 10.1006/bbrc.1996.1489

68. Guo W, Shi L, Javitch JA (2003) The fourth transmembrane segment forms the interface of the dopamine D2 receptor homodimer. *J Biol Chem* 278: 4385-4388. DOI: 10.1074/jbc.C200679200

69. Javitch JA, Shi L, Simpson MM, Chen J, Chiappa V, et al. (2000) The fourth transmembrane segment of the dopamine D2 receptor: accessibility in the binding-site crevice and position in the transmembrane bundle. *Biochemistry* 39: 12190-12199. DOI: 10.1021/bi001069m

70. Filizola M, Olmea O, Weinstein H (2002) Prediction of heterodimerization interfaces of G-protein coupled receptors with a new subtractive correlated mutation method. *Protein Eng* 15: 881-885. DOI: 10.1093/protein/15.11.881

71. Filizola M, Weinstein H (2002) Structural models for dimerization of G-protein coupled receptors: the opioid receptor homodimers. *Biopolymers* 66: 317-325. DOI: 10.1002/bip.10311

72. Johnston JM, Aburi M, Provasi D, Bortolato A, Urizar E, et al. (2011) Making structural sense of dimerization interfaces of delta opioid receptor homodimers. *Biochemistry* 50: 1682-1690. DOI: 10.1021/bi101474v

73. Luttrell LM, Lefkowitz RJ (2002) The role of beta-arrestins in the termination and transduction of G-protein-coupled receptor signals. *J Cell Sci* 115: 455-465.

74. Violin JD, Lefkowitz RJ (2007) Beta-arrestin-biased ligands at seven-transmembrane receptors. *Trends Pharmacol Sci* 28: 416-422. DOI: 10.1016/j.tips.2007.06.006

75. Savarese TM, Fraser CM (1992) In vitro mutagenesis and the search for structure-function relationships among G protein-coupled receptors. *Biochem J* 283: 1-19.

76. Herve D (2011) Identification of a specific assembly of the G protein Golf as a critical and regulated module of dopamine and adenosine-activated cAMP pathways in the striatum. *Front Neuroanat* 5: 48. DOI: 10.3389/fnana.2011.00048

77. Downes GB, Gautam N (1999) The G protein subunit gene families. Genomics 62: 544-552. DOI: 10.1006/geno.1999.5992

78. Wang D, Tan YC, Kreitzer GE, Nakai Y, Shan D, et al. (2006) G proteins G12 and G13 control the dynamic turnover of growth factor-induced dorsal ruffles. *J Biol Chem* 281: 32660-32667. DOI: 10.1074/jbc.M604588200

79. Rodbell M, Krans HM, Pohl SL, Birnbaumer L (1971) The glucagon-sensitive adenyl cyclase system in plasma membranes of rat liver. 3. Binding of glucagon: method of assay and specificity. *J Biol Chem* 246: 1861-1871.

80. Northup JK, Sternweis PC, Smigel MD, Schleifer LS, Ross EM, et al. (1980) Purification of the regulatory component of adenylate cyclase. *Proc Natl Acad Sci USA* 77: 6516-6520. DOI: 10.1073/pnas.77.11.6516

81. Milligan G, Kostenis E (2006) Heterotrimeric G-proteins: a short history. *Br J Pharmacol* 147 Suppl 1: S46-55. DOI: 10.1038/sj.bjp.0706405

82. Jones DT, Masters SB, Bourne HR, Reed RR (1990) Biochemical characterization of three stimulatory GTP-binding proteins. The large and small forms of Gs and the olfactory-specific G-protein, Golf. *J Biol Chem* 265: 2671-2676.

83. Seifert R, Wenzel-Seifert K, Lee TW, Gether U, Sanders-Bush E, et al. (1998) Different effects of Gsalpha splice variants on beta2-adrenoreceptor-mediated signaling. The beta2-adrenoreceptor coupled to the long splice variant of Gsalpha has properties of a constitutively active receptor. *J Biol Chem* 273: 5109-5116. DOI: 10.1074/jbc.273.9.5109

84. Bray P, Carter A, Simons C, Guo V, Puckett C, et al. (1986) Human cDNA clones for four species of G alpha s signal transduction protein. *Proc Natl Acad Sci USA* 83: 8893-8897. DOI: 10.1073/pnas.83.23.8893

85. Kozasa T, Itoh H, Tsukamoto T, Kaziro Y (1988) Isolation and characterization of the human Gs alpha gene. *Proc Natl Acad Sci USA* 85: 2081-2085. DOI: 10.1073/pnas.85.7.2081

86. Weinstein LS, Chen M, Liu J (2002) Gs(alpha) mutations and imprinting defects in human disease. *Ann N Y Acad Sci* 968: 173-197. DOI: 10.1111/j.1749-6632.2002.tb04335.x

87. Corradi JP, Ravyn V, Robbins AK, Hagan KW, Peters MF, et al. (2005) Alternative transcripts and evidence of imprinting of GNAL on 18p11.2. *Mol Psychiatry* 10: 1017-1025. DOI: 10.1038/sj.mp.4001713

88. Spiegel AM, Shenker A, Weinstein LS (1992) Receptor-effector coupling by G proteins: implications for normal and abnormal signal transduction. *Endocr Rev* 13: 536-565.

89. Degtyarev MY, Spiegel AM, Jones TL (1993) The G protein alpha s subunit incorporates [3H] palmitic acid and mutation of cysteine-3 prevents this modification. *Biochemistry* 32: 8057-8061. DOI: 10.1021/bi00083a001

90. Linder ME, Middleton P, Hepler JR, Taussig R, Gilman AG, et al. (1993) Lipid modifications of G proteins: alpha subunits are palmitoylated. *Proc Natl Acad Sci USA* 90: 3675-3679. DOI: 10.1073/pnas.90.8.3675

91. Mumby SM, Kleuss C, Gilman AG (1994) Receptor regulation of G-protein palmitoylation. *Proc Natl Acad Sci USA* 91: 2800-2804. DOI: 10.1073/pnas.91.7.2800

92. Allen JA, Yu JZ, Donati RJ, Rasenick MM (2005) Beta-adrenergic receptor stimulation promotes G alpha s internalization through lipid rafts: a study in living cells. *Mol Pharmacol* 67: 1493-1504. DOI: 10.1124/mol.104.008342

93. Huang C, Duncan JA, Gilman AG, Mumby SM (1999) Persistent membrane association of activated and depalmitoylated G protein alpha subunits. *Proc Natl Acad Sci USA* 96: 412-417. DOI: 10.1073/pnas.96.2.412

94. Wedegaertner PB, Bourne HR, von Zastrow M (1996) Activation-induced subcellular redistribution of Gs alpha. *Mol Biol Cell* 7: 1225-1233.

95. Yu JZ, Rasenick MM (2002) Real-time visualization of a fluorescent G(alpha)(s): dissociation of the activated G protein from plasma membrane. *Mol Pharmacol* 61: 352-359. DOI: 10.1124/mol.61.2.352

96. Zheng B, Ma YC, Ostrom RS, Lavoie C, Gill GN, et al. (2001) RGS-PX1, a GAP for GalphaS and sorting nexin in vesicular trafficking. *Science* 294: 1939-1942. DOI: 10.1126/science.1064757

97. Scholich K, Mullenix JB, Wittpoth C, Poppleton HM, Pierre SC, et al. (1999) Facilitation of signal onset and termination by adenylyl cyclase. *Science* 283: 1328-1331. DOI: 10.1126/science.283.5406.1328

98. Colombo MI, Mayorga LS, Nishimoto I, Ross EM, Stahl PD (1994) Gs regulation of endosome fusion suggests a role for signal transduction pathways in endocytosis. *J Biol Chem* 269: 14919-14923.

99. Zheng B, Lavoie C, Tang TD, Ma P, Meerloo T, et al. (2004) Regulation of epidermal growth factor receptor degradation by heterotrimeric Galphas protein. *Mol Biol Cell* 15: 5538-5550. DOI: 10.1091/mbc.E04-06-0446

100. Hynes TR, Mervine SM, Yost EA, Sabo JL, Berlot CH (2004) Live cell imaging of Gs and the beta2-adrenergic receptor demonstrates that both alphas and beta1gamma7 internalize upon stimulation and exhibit similar trafficking patterns that differ from that of the beta2-adrenergic receptor. *J Biol Chem* 279: 44101-44112. DOI: 10.1074/jbc.M405151200

101. Nagai Y, Nishimura A, Tago K, Mizuno N, Itoh H (2010) Ric-8B stabilizes the alpha subunit of stimulatory G protein by inhibiting its ubiquitination. *J Biol Chem* 285: 11114-11120. DOI: 10.1074/jbc.M109.063313

102. Mixon MB, Lee E, Coleman DE, Berghuis AM, Gilman AG, et al. (1995) Tertiary and quaternary structural changes in Gi alpha 1 induced by GTP hydrolysis. *Science* 270: 954-960. DOI: 10.1126/science.270.5238.954

103. Warner DR, Weng G, Yu S, Matalon R, Weinstein LS (1998) A novel mutation in the switch 3 region of Gsalpha in a patient with Albright hereditary osteodystrophy impairs GDP binding and receptor activation. *J Biol Chem* 273: 23976-23983. DOI: 10.1074/jbc.273.37.23976

104. Sunahara RK, Tesmer JJ, Gilman AG, Sprang SR (1997) Crystal structure of the adenylyl cyclase activator Gsalpha. *Science* 278: 1943-1947. DOI: 10.1126/science.278.5345.1943

105. Tesmer JJ, Sunahara RK, Gilman AG, Sprang SR (1997) Crystal structure of the catalytic domains of adenylyl cyclase in a complex with Gsalpha.GTPgammaS. *Science* 278: 1907-1916. DOI: 10.1126/science.278.5345.1907

106. Oldham WM, Hamm HE (2008) Heterotrimeric G protein activation by G-protein-coupled receptors. *Nat Rev Mol Cell Biol* 9: 60-71. DOI: 10.1038/nrm2299

107. Lambright DG, Noel JP, Hamm HE, Sigler PB (1994) Structural determinants for activation of the alpha-subunit of a heterotrimeric G protein. *Nature* 369: 621-628. DOI: 10.1038/369621a0

108. Noel JP, Hamm HE, Sigler PB (1993) The 2.2 A crystal structure of transducin-alpha complexed with GTP gamma S. *Nature* 366: 654-663. DOI: 10.1038/366654a0

109. Lambright DG, Sondek J, Bohm A, Skiba NP, Hamm HE, et al. (1996) The 2.0 A crystal structure of a heterotrimeric G protein. *Nature* 379: 311-319. DOI: 10.1038/379311a0

110. Kimple RJ, Kimple ME, Betts L, Sondek J, Siderovski DP (2002) Structural determinants for GoLoco-induced inhibition of nucleotide release by Galpha subunits. *Nature* 416: 878-881. DOI: 10.1038/416878a

111. Johnston CA, Willard FS, Jezyk MR, Fredericks Z, Bodor ET, et al. (2005) Structure of Galpha(i1) bound to a GDP-selective peptide provides insight into guanine nucleotide exchange. *Structure* 13: 1069-1080. DOI: 10.1016/j.str.2005.04.007

112. Johnston CA, Lobanova ES, Shavkunov AS, Low J, Ramer JK, et al. (2006) Minimal determinants for binding activated G alpha from the structure of a G alpha(i1)-peptide dimer. *Biochemistry* 45: 11390-11400. DOI: 10.1021/bi0613832

113. Warner DR, Romanowski R, Yu S, Weinstein LS (1999) Mutagenesis of the conserved residue Glu259 of Gsalpha demonstrates the importance of interactions between switches 2 and 3 for activation. *J Biol Chem* 274: 4977-4984. DOI: 10.1074/jbc.274.8.4977

114. Grishina G, Berlot CH (1998) Mutations at the domain interface of GSalpha impair receptor-mediated activation by altering receptor and guanine nucleotide binding. *J Biol Chem* 273: 15053-15060. DOI: 10.1074/jbc.273.24.15053

115. Marsh SR, Grishina G, Wilson PT, Berlot CH (1998) Receptor-mediated activation of Gsalpha: evidence for intramolecular signal transduction. *Mol Pharmacol* 53: 981-990.

116. Iiri T, Farfel Z, Bourne HR (1997) Conditional activation defect of a human Gsalpha mutant. *Proc Natl Acad Sci USA* 94: 5656-5661. DOI: 10.1073/pnas.94.11.5656

117. Li Q, Cerione RA (1997) Communication between switch II and switch III of the transducin alpha subunit is essential for target activation. *J Biol Chem* 272: 21673-21676. DOI: 10.1074/jbc.272.35.21673

118. Warner DR, Weinstein LS (1999) A mutation in the heterotrimeric stimulatory guanine nucleotide binding protein alpha-subunit with impaired receptor-mediated activation because of elevated GTPase activity. *Proc Natl Acad Sci USA* 96: 4268-4272. DOI: 10.1073/pnas.96.8.4268

119. Wall MA, Coleman DE, Lee E, Iniguez-Lluhi JA, Posner BA, et al. (1995) The structure of the G protein heterotrimer Gi alpha 1 beta 1 gamma 2. *Cell* 83: 1047-1058. DOI: 10.1016/0092-8674(95)90220-1

120. Conklin BR, Herzmark P, Ishida S, Voyno-Yasenetskaya TA, Sun Y, et al. (1996) Carboxyl-terminal mutations of Gq alpha and Gs alpha that alter the fidelity of receptor activation. *Mol Pharmacol* 50: 885-890.

121. Grishina G, Berlot CH (2000) A surface-exposed region of G(salpha) in which substitutions decrease receptor-mediated activation and increase receptor affinity. *Mol Pharmacol* 57: 1081-1092.

122. Mazzoni MR, Taddei S, Giusti L, Rovero P, Galoppini C, et al. (2000) A galpha(s) carboxyl-terminal peptide prevents G(s) activation by the A(2A) adenosine receptor. *Mol Pharmacol* 58: 226-236.

123. Schwindinger WF, Miric A, Zimmerman D, Levine MA (1994) A novel Gs alpha mutant in a patient with Albright hereditary osteodystrophy uncouples cell surface receptors from adenylyl cyclase. *J Biol Chem* 269: 25387-25391.

124. Simonds WF, Goldsmith PK, Woodard CJ, Unson CG, Spiegel AM (1989) Receptor and effector interactions of Gs. Functional studies with antibodies to the alpha s carboxyl-terminal decapeptide. *FEBS Lett* 249: 189-194. DOI: 10.1016/0014-5793(89)80622-2

125. Sullivan KA, Miller RT, Masters SB, Beiderman B, Heideman W, et al. (1987) Identification of receptor contact site involved in receptor-G protein coupling. *Nature* 330: 758-760. DOI: 10.1038/330758a0

126. Hamm HE, Kaya AI, Gilbert JA, 3rd, Preininger AM (2013) Linking receptor activation to changes in Sw I and II of Galpha proteins. *J Struct Biol* (Epub ahead of print). DOI: 10.1016/j.jsb.2013.02.016

127. Ui M, Katada T (1990) Bacterial toxins as probe for receptor-Gi coupling. A*dv Second Messenger Phosphoprotein Res* 24: 63-69.

128. Sternweis PC, Robishaw JD (1984) Isolation of two proteins with high affinity for guanine nucleotides from membranes of bovine brain. *J Biol Chem* 259: 13806-13813.

129. Neer EJ, Lok JM, Wolf LG (1984) Purification and properties of the inhibitory guanine nucleotide regulatory unit of brain adenylate cyclase. *J Biol Chem* 259: 14222-14229.

130. Olate J, Allende JE (1991) Structure and function of G proteins. *Pharmacol Ther* 51: 403-419. DOI: 10.1016/0163-7258(91)90068-W

131. Spiegel AM, Gierschik P, Levine MA, Downs RW, Jr. (1985) Clinical implications of guanine nucleotide-binding proteins as receptor-effector couplers. *N Engl J Med* 312: 26-33. DOI: 10.1056/NEJM198501033120106

132. Offermanns S (2003) G-proteins as transducers in transmembrane signalling. *Prog Biophys Mol Biol* 83: 101-130. DOI: 10.1016/S0079-6107(03)00052-X

133. Albarran-Juarez J, Gilsbach R, Piekorz RP, Pexa K, Beetz N, et al. (2009) Modulation of alpha2-adrenoceptor functions by heterotrimeric Galphai protein isoforms. *J Pharmacol Exp* Ther 331: 35-44. DOI: 10.1124/jpet.109.157230

134. Wedegaertner PB, Wilson PT, Bourne HR (1995) Lipid modifications of trimeric G proteins. *J Biol Chem* 270: 503-506. DOI: 10.1074/jbc.270.2.503

135. Jiang M, Gold MS, Boulay G, Spicher K, Peyton M, et al. (1998) Multiple neurological abnormalities in mice deficient in the G protein Go. *Proc Natl Acad Sci USA* 95: 3269-3274. DOI: 10.1073/pnas.95.6.3269

136. Strathmann M, Wilkie TM, Simon MI (1990) Alternative splicing produces transcripts encoding two forms of the alpha subunit of GTP-binding protein Go. P*roc Natl Acad Sci USA* 87: 6477-6481. DOI: 10.1073/pnas.87.17.6477

137. Hsu WH, Rudolph U, Sanford J, Bertrand P, Olate J, et al. (1990) Molecular cloning of a novel splice variant of the alpha subunit of the mammalian Go protein. *J Biol Chem* 265: 11220-11226.

138. Exner T, Jensen ON, Mann M, Kleuss C, Nurnberg B (1999) Posttranslational modification of Galphao1 generates Galphao3, an abundant G protein in brain. *Proc Natl Acad Sci USA* 96: 1327-1332. DOI: 10.1073/pnas.96.4.1327

139. Cerione RA, Regan JW, Nakata H, Codina J, Benovic JL, et al. (1986) Functional reconstitution of the alpha 2-adrenergic receptor with guanine nucleotide regulatory proteins in phospholipid vesicles. *J Biol Chem* 261: 3901-3909.

140. Florio VA, Sternweis PC (1985) Reconstitution of resolved muscarinic cholinergic receptors with purified GTP-binding proteins. *J Biol Chem* 260: 3477-3483.

141. Hescheler J, Rosenthal W, Trautwein W, Schultz G (1987) The GTP-binding protein, Go, regulates neuronal calcium channels. *Nature* 325: 445-447. DOI: 10.1038/325445a0

142. Kleuss C, Hescheler J, Ewel C, Rosenthal W, Schultz G, et al. (1991) Assignment of G-protein subtypes to specific receptors inducing inhibition of calcium currents. *Nature* 353: 43-48. DOI: /10.1038/353043a0

143. Taussig R, Sanchez S, Rifo M, Gilman AG, Belardetti F (1992) Inhibition of the omega-conotoxin-sensitive calcium current by distinct G proteins. *Neuron* 8: 799-809. DOI: 10.1016/0896-6273(92)90100-R

144. Chen C, Clarke IJ (1996) G(o)-2 protein mediates the reduction in Ca2+ currents by somatostatin in cultured ovine somatotrophs. *J Physiol* 491: 21-29.

145. Strittmatter SM, Valenzuela D, Kennedy TE, Neer EJ, Fishman MC (1990) G0 is a major growth cone protein subject to regulation by GAP-43. *Nature* 344: 836-841. DOI: 10.1038/344836a0

146. Nishimoto I, Okamoto T, Matsuura Y, Takahashi S, Okamoto T, et al. (1993) Alzheimer amyloid protein precursor complexes with brain GTP-binding protein G(o). *Nature* 362: 75-79. DOI: 10.1038/362075a0

147. Jordan JD, Iyengar R (2002) Identification of putative direct effectors for G alpha o, using yeast two-hybrid method. *Methods Enzymol* 345: 140-149. DOI: 10.1016/S0076-6879(02)45013-6

148. Jordan JD, Carey KD, Stork PJ, Iyengar R (1999) Modulation of rap activity by direct interaction of Galpha(o) with Rap1 GTPase-activating protein. *J Biol Chem* 274: 21507-21510. DOI: 10.1074/jbc.274.31.21507

149. Gangal M, Clifford T, Deich J, Cheng X, Taylor SS, et al. (1999) Mobilization of the A-kinase N-myristate through an isoform-specific intermolecular switch. *Proc Natl Acad Sci USA* 96: 12394-12399. DOI: 10.1073/pnas.96.22.12394

150. Fong HK, Yoshimoto KK, Eversole-Cire P, Simon MI (1988) Identification of a GTP-binding protein alpha subunit that lacks an apparent ADP-ribosylation site for pertussis toxin. *Proc Natl Acad Sci USA* 85: 3066-3070. DOI: 10.1073/pnas.85.9.3066

151. Matsuoka M, Itoh H, Kozasa T, Kaziro Y (1988) Sequence analysis of cDNA and genomic DNA for a putative pertussis toxin-insensitive guanine nucleotide-binding regulatory protein alpha subunit. *Proc Natl Acad Sci USA* 85: 5384-5388. DOI: 10.1073/pnas.85.15.5384

152. Wong YH, Conklin BR, Bourne HR (1992) Gz-mediated hormonal inhibition of cyclic AMP accumulation. *Science* 255: 339-342. DOI: 10.1126/science.1347957

153. Jeong SW, Ikeda SR (1998) G protein alpha subunit G alpha z couples neurotransmitter receptors to ion channels in sympathetic neurons. *Neuron* 21: 1201-1212. DOI: 10.1016/S0896-6273(00)80636-4

154. Lounsbury KM, Casey PJ, Brass LF, Manning DR (1991) Phosphorylation of Gz in human platelets. Selectivity and site of modification. *J Biol Chem 266*: 22051-22056.

155. Mumby SM, Heukeroth RO, Gordon JI, Gilman AG (1990) G-protein alpha-subunit expression, myristoylation, and membrane association in COS cells. *Proc Natl Acad Sci USA* 87: 728-732. DOI: 10.1073/pnas.87.2.728

156. Hallak H, Muszbek L, Laposata M, Belmonte E, Brass LF, et al. (1994) Covalent binding of arachidonate to G protein alpha subunits of human platelets. *J Biol Chem* 269: 4713-4716.

157. Casey PJ, Fong HK, Simon MI, Gilman AG (1990) Gz, a guanine nucleotide-binding protein with unique biochemical properties. *J Biol Chem* 265: 2383-2390.

158. Fields TA, Casey PJ (1995) Phosphorylation of Gz alpha by protein kinase C blocks interaction with the beta gamma complex. *J Biol Chem* 270: 23119-23125. DOI: 10.1074/jbc.270.39.23119

159. Wang J, Frost JA, Cobb MH, Ross EM (1999) Reciprocal signaling between heterotrimeric G proteins and the p21-stimulated protein kinase. *J Biol Chem* 274: 31641-31647. DOI: 10.1074/jbc.274.44.31641

160. Ho MK, Wong YH (2001) G(z) signaling: emerging divergence from G(i) signaling. *Oncogene* 20: 1615-1625. DOI: 10.1038/sj.onc.1204190

161. Exton JH (1996) Regulation of phosphoinositide phospholipases by hormones, neurotransmitters, and other agonists linked to G proteins. *Annu Rev Pharmacol Toxicol* 36: 481-509. DOI: 10.1146/annurev.pa.36.040196.002405

162. Rebecchi MJ, Pentyala SN (2000) Structure, function, and control of phosphoinositide-specific phospholipase *C. Physiol Rev* 80: 1291-1335.

163. Rhee SG (2001) Regulation of phosphoinositide-specific phospholipase *C. Annu Rev Biochem* 70: 281-312. DOI: 10.1146/annurev.biochem.70.1.281

164. Wilkie TM, Scherle PA, Strathmann MP, Slepak VZ, Simon MI (1991) Characterization of G-protein alpha subunits in the Gq class: expression in murine tissues and in stromal and hematopoietic cell lines. *Proc Natl Acad Sci USA* 88: 10049-10053. DOI: 10.1073/pnas.88.22.10049

165. Amatruda TT, 3rd, Steele DA, Slepak VZ, Simon MI (1991) G alpha 16, a G protein alpha subunit specifically expressed in hematopoietic cells. *Proc Natl Acad Sci USA* 88: 5587-5591. DOI: 10.1073/pnas.88.13.5587

166. Fukami K, Inanobe S, Kanemaru K, Nakamura Y (2010) Phospholipase C is a key enzyme regulating intracellular calcium and modulating the phosphoinositide balance. *Prog Lipid Res* 49: 429-437. DOI: 10.1016/j.plipres.2010.06.001

167. Gresset A, Sondek J, Harden TK (2012) The phospholipase C isozymes and their regulation. *Subcell Biochem* 58: 61-94. DOI: 10.1007/978-94-007-3012-0_3

168. Clapham DE, Neer EJ (1997) G protein beta gamma subunits. *Annu Rev Pharmacol Toxicol* 37: 167-203. DOI: 10.1146/annurev.pharmtox.37.1.167

169. Gautam N, Downes GB, Yan K, Kisselev O (1998) The G-protein betagamma complex. *Cell Signal* 10: 447-455. DOI: 10.1016/S0898-6568(98)00006-0

170. Schwindinger WF, Robishaw JD (2001) Heterotrimeric G-protein betagamma-dimers in growth and differentiation. *Oncogene* 20: 1653-1660. DOI: 10.1038/sj.onc.1204181

171. Dupre DJ, Robitaille M, Rebois RV, Hebert TE (2009) The role of Gbetagamma subunits in the organization, assembly, and function of GPCR signaling complexes. *Annu Rev Pharmacol Toxicol* 49: 31-56. DOI: 10.1146/annurev-pharmtox-061008-103038

172. Yan K, Kalyanaraman V, Gautam N (1996) Differential ability to form the G protein betagamma complex among members of the beta and gamma subunit families. *J Biol Chem*271: 7141-7146. DOI: 10.1074/jbc.271.12.7141

173. Wang Q, Mullah BK, Robishaw JD (1999) Ribozyme approach identifies a functional association between the G protein beta1gamma7 subunits in the beta-adrenergic receptor signaling pathway. *J Biol Chem* 274: 17365-17371. DOI: 10.1074/jbc.274.24.17365

174. Asano T, Morishita R, Ueda H, Kato K (1999) Selective association of G protein beta(4) with gamma(5) and gamma(12) subunits in bovine tissues. *J Biol Chem* 274: 21425-21429. DOI: 10.1074/jbc.274.30.21425

175. Sondek J, Bohm A, Lambright DG, Hamm HE, Sigler PB (1996) Crystal structure of a G-protein beta gamma dimer at 2.1A resolution. *Nature* 379: 369-374. DOI: 10.1038/379369a0

176. Logothetis DE, Kurachi Y, Galper J, Neer EJ, Clapham DE (1987) The beta gamma subunits of GTP-binding proteins activate the muscarinic K+ channel in heart. *Nature* 325: 321-326. DOI: 10.1038/325321a0

177. Taussig R, Gilman AG (1995) Mammalian membrane-bound adenylyl cyclases. *J Biol Chem* 270: 1-4. DOI: 10.1074/jbc.270.1.1

178. Defer N, Best-Belpomme M, Hanoune J (2000) Tissue specificity and physiological relevance of various isoforms of adenylyl cyclase. *Am J Physiol Renal Physiol* 279: F400-416.

179. Hanoune J, Defer N (2001) Regulation and role of adenylyl cyclase isoforms. *Annu Rev Pharmacol Toxicol* 41: 145-174. DOI: 10.1146/annurev.pharmtox.41.1.145

180. Sunahara RK, Taussig R (2002) Isoforms of mammalian adenylyl cyclase: multiplicities of signaling. *Mol Interv* 2: 168-184. DOI: 10.1124/mi.2.3.168

181. Kamenetsky M, Middelhaufe S, Bank EM, Levin LR, Buck J, et al. (2006) Molecular details of cAMP generation in mammalian cells: a tale of two systems. *J Mol Biol* 362: 623-639. DOI: 10.1016/j.jmb.2006.07.045

182. Patel TB, Du Z, Pierre S, Cartin L, Scholich K (2001) Molecular biological approaches to unravel adenylyl cyclase signaling and function. *Gene* 269: 13-25. DOI: 10.1016/S0378-1119(01)00448-6

183. Buck J, Sinclair ML, Schapal L, Cann MJ, Levin LR (1999) Cytosolic adenylyl cyclase defines a unique signaling molecule in mammals. *Proc Natl Acad Sci USA* 96: 79-84. DOI: 10.1073/pnas.96.1.79

184. Chen Y, Cann MJ, Litvin TN, Iourgenko V, Sinclair ML, et al. (2000) Soluble adenylyl cyclase as an evolutionarily conserved bicarbonate sensor. *Science* 289: 625-628. DOI: 10.1126/science.289.5479.625

185. Jaiswal BS, Conti M (2003) Calcium regulation of the soluble adenylyl cyclase expressed in mammalian spermatozoa. *Proc Natl Acad Sci USA* 100: 10676-10681. DOI: 10.1073/pnas.1831008100

186. Litvin TN, Kamenetsky M, Zarifyan A, Buck J, Levin LR (2003) Kinetic properties of "soluble" adenylyl cyclase - Synergism between calcium and bicarbonate. *J Biol Chem* 278: 15922-15926. DOI: 10.1074/jbc.M212475200

187. Thomas JM, Hoffman BB (1996) Isoform-specific sensitization of adenylyl cyclase activity by prior activation of inhibitory receptors: role of beta gamma subunits in transducing enhanced activity of the type VI isoform. *Mol Pharmacol* 49: 907-914.

188. Harry A, Chen Y, Magnusson R, Iyengar R, Weng G (1997) Differential regulation of adenylyl cyclases by Galphas. *J Biol Chem* 272: 19017-19021. DOI: 10.1074/jbc.272.30.19017

189. Wang T, Brown MJ (2004) Differential expression of adenylyl cyclase subtypes in human cardiovascular system. *Mol Cell Endocrinol* 223: 55-62. DOI: 10.1016/j.mce.2004.05.012

190. Pierre S, Eschenhagen T, Geisslinger G, Scholich K (2009) Capturing adenylyl cyclases as potential drug targets. *Nat Rev Drug Discov* 8: 321-335. DOI: 10.1038/nrd2827

191. Pavan B, Biondi C, Dalpiaz A (2009) Adenylyl cyclases as innovative therapeutic goals. *Drug Discov Today* 14: 982-991. DOI: 10.1016/j.drudis.2009.07.007

192. Sunahara RK, Dessauer CW, Gilman AG (1996) Complexity and diversity of mammalian adenylyl cyclases. *Annu Rev Pharmacol Toxicol* 36: 461-480. DOI: 10.1146/annurev.pa.36.040196.002333

193. Smit MJ, Iyengar R (1998) Mammalian adenylyl cyclases. *Adv Second Messenger Phosphoprotein Res* 32: 1-21. DOI: 10.1016/S1040-7952(98)80003-7

194. Premont RT, Macrae AD, Stoffel RH, Chung N, Pitcher JA, et al. (1996) Characterization of the G protein-coupled receptor kinase GRK4. Identification of four splice variants. *J Biol Chem* 271: 6403-6410. DOI: 10.1074/jbc.271.11.6403

195. Yan SZ, Huang ZH, Andrews RK, Tang WJ (1998) Conversion of forskolin-insensitive to forskolin-sensitive (mouse-type IX) adenylyl cyclase. *Mol Pharmacol* 53: 182-187.

196. Tang WJ, Krupinski J, Gilman AG (1991) Expression and characterization of calmodulin-activated (type I) adenylylcyclase. *J Biol Chem* 266: 8595-8603.

197. Krupinski J, Lehman TC, Frankenfield CD, Zwaagstra JC, Watson PA (1992) Molecular diversity in the adenylylcyclase family. Evidence for eight forms of the enzyme and cloning of type VI. *J Biol Chem* 267: 24858-24862.

198. Cali JJ, Parekh RS, Krupinski J (1996) Splice variants of type VIII adenylyl cyclase. Differences in glycosylation and regulation by Ca2+/calmodulin. *J Biol Chem* 271: 1089-1095. DOI: 10.1074/jbc.271.2.1089

199. Feinstein PG, Schrader KA, Bakalyar HA, Tang WJ, Krupinski J, et al. (1991) Molecular cloning and characterization of a Ca2+/calmodulin-insensitive adenylyl cyclase from rat brain. *Proc Natl Acad Sci USA* 88: 10173-10177. DOI: 10.1073/pnas.88.22.10173

200. Gao BN, Gilman AG (1991) Cloning and expression of a widely distributed (type IV) adenylyl cyclase. *Proc Natl Acad Sci USA* 88: 10178-10182. DOI: 10.1073/pnas.88.22.10178

201. Tang WJ, Gilman AG (1991) Type-specific regulation of adenylyl cyclase by G protein beta gamma subunits. *Science* 254: 1500-1503. DOI: 10.1126/science.1962211

202. Yoshimura M, Ikeda H, Tabakoff B (1996) mu-Opioid receptors inhibit dopamine-stimulated activity of type V adenylyl cyclase but enhance dopamine-stimulated activity of type VII adenylyl cyclase. *Mol Pharmacol* 50: 43-51.

203. Johnson RA, Shoshani I (1990) Kinetics of "P"-site-mediated inhibition of adenylyl cyclase and the requirements for substrate. *J Biol Chem* 265: 11595-11600.

204. Dessauer CW, Gilman AG (1997) The catalytic mechanism of mammalian adenylyl cyclase. Equilibrium binding and kinetic analysis of P-site inhibition. *J Biol Chem* 272: 27787-27795.

205. Hurley JH (1999) Structure, mechanism, and regulation of mammalian adenylyl cyclase. *J Biol Chem* 274: 7599-7602. DOI: 10.1074/jbc.274.12.7599

206. Offermanns S, Toombs CF, Hu YH, Simon MI (1997) Defective platelet activation in G alpha(q)-deficient mice. *Nature* 389: 183-186. DOI: 10.1038/38284

207. Kelley GG, Reks SE, Smrcka AV (2004) Hormonal regulation of phospholipase Cepsilon through distinct and overlapping pathways involving G12 and Ras family G-proteins. *Biochem J* 378: 129-139. DOI: 10.1042/BJ20031370

208. Lopez I, Mak EC, Ding J, Hamm HE, Lomasney JW (2001) A novel bifunctional phospholipase c that is regulated by Galpha 12 and stimulates the Ras/mitogen-activated protein kinase pathway. *J Biol Chem* 276: 2758-2765. DOI: 10.1074/jbc.M008119200

209. Wing MR, Houston D, Kelley GG, Der CJ, Siderovski DP, et al. (2001) Activation of phospholipase C-epsilon by heterotrimeric G protein betagamma-subunits. *J Biol Chem* 276: 48257-48261.

210. Streb H, Irvine RF, Berridge MJ, Schulz I (1983) Release of Ca2+ from a nonmitochondrial intracellular store in pancreatic acinar cells by inositol-1,4,5-trisphosphate. *Nature* 306: 67-69. DOI: 10.1038/306067a0

211. Cockcroft S (2006) The latest phospholipase C, PLC eta, is implicated in neuronal function. *Trends Biochem Sci* 31: 4-7. DOI: 10.1016/j.tibs.2005.11.003

212. Essen LO, Perisic O, Cheung R, Katan M, Williams RL (1996) Crystal structure of a mammalian phosphoinositide-specific phospholipase C delta. *Nature* 380: 595-602. DOI: 10.1038/380595a0

213. Wierenga RK (2001) The TIM-barrel fold: a versatile framework for efficient enzymes. *FEBS Lett* 492: 193-198. DOI: 10.1016/S0014-5793(01)02236-0

214. Nalefski EA, Falke JJ (1996) The C2 domain calcium-binding motif: structural and functional diversity. *Protein Sci* 5: 2375-2390. DOI: 10.1002/pro.5560051201

215. Rhee SG, Choi KD (1992) Regulation of inositol phospholipid-specific phospholipase C isozymes. *J Biol Chem* 267: 12393-12396.

216. Suh PG, Park JI, Manzoli L, Cocco L, Peak JC, et al. (2008) Multiple roles of phosphoinositide-specific phospholipase C isozymes. *BMB Rep* 41: 415-434. DOI: 10.5483/BMBRep.2008.41.6.415

217. Fukami K (2002) Structure, regulation, and function of phospholipase C isozymes. *J Biochem* 131: 293-299. DOI: 10.1093/oxfordjournals.jbchem.a003102

218. Waldo GL, Ricks TK, Hicks SN, Cheever ML, Kawano T, et al. (2010) Kinetic scaffolding mediated by a phospholipase C-beta and Gq signaling complex. *Science* 330: 974-980. DOI: 10.1126/science.1193438

219. Singer WD, Brown HA, Sternweis PC (1997) Regulation of eukaryotic phosphatidylinositol-specific phospholipase C and phospholipase D. *Annu Rev Biochem* 66: 475-509. DOI: 10.1146/annurev.biochem.66.1.475

220. Hicks SN, Jezyk MR, Gershburg S, Seifert JP, Harden TK, et al. (2008) General and versatile autoinhibition of PLC isozymes. *Mol Cell* 31: 383-394. DOI: 10.1016/j.molcel.2008.06.018

221. Jezyk MR, Snyder JT, Gershberg S, Worthylake DK, Harden TK, et al. (2006) Crystal structure of Rac1 bound to its effector phospholipase C-beta 2. *Nat Struct Mol Biol* 13: 1135-1140. DOI: 10.1038/nsmb1175

222. Ji QS, Winnier GE, Niswender KD, Horstman D, Wisdom R, et al. (1997) Essential role of the tyrosine kinase substrate phospholipase C-gamma1 in mammalian growth and development. *Proc Natl Acad Sci USA* 94: 2999-3003. DOI: 10.1073/pnas.94.7.2999

223. Liao HJ, Kume T, McKay C, Xu MJ, Ihle JN, et al. (2002) Absence of erythrogenesis and vasculogenesis in Plcg1-deficient mice. *J Biol Chem* 277: 9335-9341. DOI: 10.1074/jbc.M109955200

224. Jakus Z, Simon E, Frommhold D, Sperandio M, Mocsai A (2009) Critical role of phospholipase Cgamma2 in integrin and Fc receptor-mediated neutrophil functions and the effector phase of autoimmune arthritis. *J Exp Med* 206: 577-593. DOI: 10.1084/jem.20081859

225. Wang D, Feng J, Wen R, Marine JC, Sangster MY, et al. (2000) Phospholipase Cgamma2 is essential in the functions of B cell and several Fc receptors. *Immunity* 13: 25-35. DOI: 10.1016/S1074-7613(00)00005-4

226. Katan M, Williams RL (1997) Phosphoinositide-specific phospholipase C: structural basis for catalysis and regulatory interactions. *Semin Cell Dev Biol* 8: 287-296. DOI: 10.1006/scdb.1997.0150

227. Pawson T, Gish GD (1992) SH2 and SH3 domains: from structure to function. *Cell* 71: 359-362. DOI: 10.1016/0092-8674(92)90504-6

228. Choi JH, Ryu SH, Suh PG (2007) On/off-regulation of phospholipase C-gamma 1-mediated signal transduction. *Adv Enzyme Regul* 47: 104-116. DOI: 10.1016/j.advenzreg.2006.12.010

229. Falasca M, Logan SK, Lehto VP, Baccante G, Lemmon MA, et al. (1998) Activation of phospholipase C gamma by PI 3-kinase-induced PH domain-mediated membrane targeting. *EMBO J* 17: 414-422. DOI: 10.1093/emboj/17.2.414

230. Singh SM, Murray D (2003) Molecular modeling of the membrane targeting of phospholipase C pleckstrin homology domains. *Protein Sci* 12: 1934-1953. DOI: 10.1110/ps.0358803

231. Ferguson KM, Lemmon MA, Schlessinger J, Sigler PB (1995) Structure of the high affinity complex of inositol trisphosphate with a phospholipase C pleckstrin homology domain. *Cell* 83: 1037-1046. DOI: 10.1016/0092-8674(95)90219-8

232. Essen LO, Perisic O, Lynch DE, Katan M, Williams RL (1997) A ternary metal binding site in the C2 domain of phosphoinositide-specific phospholipase C-delta1. *Biochem* 36: 2753-2762. DOI: 10.1021/bi962466t

233. Kelley GG, Reks SE, Ondrako JM, Smrcka AV (2001) Phospholipase C epsilon: a novel Ras effector. *EMBO J* 20: 743-754. DOI: 10.1093/emboj/20.4.743

234. Song C, Hu CD, Masago M, Kariyai K, Yamawaki-Kataoka Y, et al. (2001) Regulation of a novel human phospholipase C, PLCepsilon, through membrane targeting by Ras. *J Biol Chem* 276: 2752-2757. DOI: 10.1074/jbc.M008324200

235. Kim CG, Park D, Rhee SG (1996) The role of carboxyl-terminal basic amino acids in Gq-alpha-dependent activation, particulate association, and nuclear localization of phospholipase C-beta1. *J Biol Chem* 271: 21187-21192. DOI: 10.1074/jbc.271.35.21187

236. Hwang JI, Oh YS, Shin KJ, Kim H, Ryu SH, et al. (2005) Molecular cloning and characterization of a novel phospholipase C, PLC-eta. *Biochem J* 389: 181-186. DOI: 10.1042/BJ20041677

237. Nakahara M, Shimozawa M, Nakamura Y, Irino Y, Morita M, et al. (2005) A novel phospholipase C, PLC eta 2, is a neuron-specific isozyme. *J Biol Chem* 280: 29128-29134. DOI: 10.1074/jbc.M503817200

238. Mark MD, Herlitze S (2000) G-protein mediated gating of inward-rectifier K+ channels. *Eur J Biochem* 267: 5830-5836. DOI: 10.1046/j.1432-1327.2000.01670.x

239. Zylbergold P, Ramakrishnan N, Hebert T (2010) The role of G proteins in assembly and function of Kir3 inwardly rectifying potassium channels. *Channels* 4: 411-421. DOI: 10.4161/chan.4.5.13327

240. Karschin C, Dissmann E, Stuhmer W, Karschin A (1996) IRK(1-3) and GIRK(1-4) inwardly rectifying K+ channel mRNAs are differentially expressed in the adult rat brain. *J Neurosci* 16: 3559-3570.

241. Bichet D, Haass FA, Jan LY (2003) Merging functional studies with structures of inward-rectifier K(+) channels. *Nat Rev Neurosci* 4: 957-967. DOI: 10.1038/nrn1244

242. Kuo A, Gulbis JM, Antcliff JF, Rahman T, Lowe ED, et al. (2003) Crystal structure of the potassium channel KirBac1.1 in the closed state. *Science* 300: 1922-1926. DOI: 10.1126/science.1085028

243. Heginbotham L, Lu Z, Abramson T, MacKinnon R (1994) Mutations in the K+ channel signature sequence. *Biophys J* 66: 1061-1067. DOI: 10.1016/S0006-3495(94)80887-2

244. Hibino H, Inanobe A, Furutani K, Murakami S, Findlay I, et al. (2010) Inwardly rectifying potassium channels: their structure, function, and physiological roles. *Physiol Rev* 90: 291-366. DOI: 10.1152/physrev.00021.2009

245. Glowatzki E, Fakler G, Brandle U, Rexhausen U, Zenner HP, et al. (1995) Subunit-dependent assembly of inward-rectifier K+ channels. *Proc Biol Sci* 261: 251-261. DOI: 10.1098/rspb.1995.0145

246. Yang J, Jan YN, Jan LY (1995) Determination of the subunit stoichiometry of an inwardly rectifying potassium channel. *Neuron* 15: 1441-1447. DOI: 10.1016/0896-6273(95)90021-7

247. Nishida M, Cadene M, Chait BT, MacKinnon R (2007) Crystal structure of a Kir3.1-prokaryotic Kir channel chimera. *EMBO J* 26: 4005-4015. DOI: 10.1038/sj.emboj.7601828

248. Logothetis DE, Lupyan D, Rosenhouse-Dantsker A (2007) Diverse Kir modulators act in close proximity to residues implicated in phosphoinositide binding. *J Physiol* 582: 953-965. DOI: 10.1113/jphysiol.2007.133157

249. Luscher C, Slesinger PA (2010) Emerging roles for G protein-gated inwardly rectifying potassium (GIRK) channels in health and disease. *Nat Rev Neurosci* 11: 301-315. DOI: 10.1038/nrn2834

250. Isomoto S, Kondo C, Takahashi N, Matsumoto S, Yamada M, et al. (1996) A novel ubiquitously distributed isoform of GIRK2 (GIRK2B) enhances GIRK1 expression of the G-protein-gated K+ current in Xenopus oocytes. *Biochem Biophys Res Commun* 218: 286-291. DOI: 10.1006/bbrc.1996.0050

251. Ivanina T, Rishal I, Varon D, Mullner C, Frohnwieser-Steinecke B, et al. (2003) Mapping the Gbetagamma-binding sites in GIRK1 and GIRK2 subunits of the G protein-activated K+ channel. *J Biol Chem* 278: 29174-29183. DOI: 10.1074/jbc.M304518200

252. Dascal N (2001) Ion-channel regulation by G proteins. *Trends Endocrinol Metab* 12: 391-398. DOI: 10.1016/S1043-2760(01)00475-1

253. Ivanina T, Varon D, Peleg S, Rishal I, Porozov Y, et al. (2004) Galphai1 and Galphai3 differentially interact with, and regulate, the G protein-activated K+ channel. *J Biol Chem* 279: 17260-17268. DOI: 10.1074/jbc.M313425200

254. Mirshahi T, Mittal V, Zhang H, Linder ME, Logothetis DE (2002) Distinct sites on G protein beta gamma subunits regulate different effector functions. *J Biol Chem* 277: 36345-36350. DOI: 10.1074/jbc.M205359200

255. Kurachi Y, Ishii M (2004) Cell signal control of the G protein-gated potassium channel and its subcellular localization. *J Physiol* 554: 285-294. DOI: 10.1113/jphysiol.2003.048439

256. Corey S, Clapham DE (2001) The Stoichiometry of Gbeta gamma binding to G-protein-regulated inwardly rectifying K+ channels (GIRKs). *J Biol Chem* 276: 11409-11413. DOI: 10.1074/jbc.M100058200

257. Sadja R, Alagem N, Reuveny E (2002) Graded contribution of the Gbeta gamma binding domains to GIRK channel activation. *Proc Natl Acad Sci USA* 99: 10783-10788. DOI: 10.1073/pnas.162346199

258. Huang CL, Jan YN, Jan LY (1997) Binding of the G protein betagamma subunit to multiple regions of G protein-gated inward-rectifying K+ channels. *FEBS Lett* 405: 291-298. DOI: 10.1016/S0014-5793(97)00197-X

259. Dunlap K, Fischbach GD (1978) Neurotransmitters decrease the calcium ocmponent of sensory neurone action potentials. *Nature* 276: 837-839. DOI: 10.1038/276837a0

260. Dunlap K, Fischbach GD (1981) Neurotransmitters decrease the calcium conductance activated by depolarization of embryonic chick sensory neurones. *J Physiol* 317: 519-535.

261. Holz GGt, Rane SG, Dunlap K (1986) GTP-binding proteins mediate transmitter inhibition of voltage-dependent calcium channels. *Nature* 319: 670-672. DOI: 10.1038/319670a0

262. Scott RH, Dolphin AC (1986) Regulation of calcium currents by a GTP analogue: potentiation of (-)-baclofen-mediated inhibition. *Neurosci Lett* 69: 59-64. DOI: 10.1016/0304-3940(86)90414-3

263. Dolphin AC (2006) A short history of voltage-gated calcium channels. *Br J Pharmacol* 147 Suppl 1: S56-62. DOI: 10.1038/sj.bjp.0706442

264. Nicoll RA, Tomita S, Bredt DS (2006) Auxiliary subunits assist AMPA-type glutamate receptors. *Science* 311: 1253-1256. DOI: 10.1126/science.1123339

265. Reuter H (1979) Properties of two inward membrane currents in the heart. *Annu Rev Physiol* 41: 413-424. DOI: 10.1146/annurev.ph.41.030179.002213

266. Lee CW, Eu YJ, Min HJ, Cho EM, Lee JH, et al. (2011) Expression and characterization of recombinant kurtoxin, an inhibitor of T-type voltage-gated calcium channels. *Biochem Biophys Res Commun* 416: 277-282. DOI: 10.1016/j.bbrc.2011.11.003

267. Tsien RW, Lipscombe D, Madison DV, Bley KR, Fox AP (1988) Multiple Types of Neuronal Calcium Channels and Their Selective Modulation. *Trends Neurosci* 11: 431-438. DOI: 10.1016/0166-2236(88)90194-4

268. Carbone E, Lux HD (1984) A low voltage-activated, fully inactivating Ca channel in vertebrate sensory neurones. *Nature* 310: 501-502. DOI: 10.1038/310501a0

269. Ikeda SR, Dunlap K (1999) Voltage-dependent modulation of N-type calcium channels: role of G protein subunits. *Adv Second Messenger Phosphoprotein Res* 33: 131-151. DOI: 10.1016/S1040-7952(99)80008-1

270. Kaneko S, Akaike A, Satoh M (1999) Receptor-mediated modulation of voltage-dependent Ca2+ channels via heterotrimeric G-proteins in neurons. *Jpn J Pharmacol* 81: 324-331. DOI: 10.1254/jjp.81.324

271. Ikeda SR (1996) Voltage-dependent modulation of N-type calcium channels by G-protein beta gamma subunits. *Nature* 380: 255-258. DOI: 10.1038/380255a0

272. Herlitze S, Garcia DE, Mackie K, Hille B, Scheuer T, et al. (1996) Modulation of Ca2+ channels by G-protein beta gamma subunits. *Nature* 380: 258-262. DOI: 10.1038/380258a0

273. Zhong JM, Hume JR, Keef KD (2001) beta-Adrenergic receptor stimulation of L-type Ca2+ channels in rabbit portal vein myocytes involves both alpha s and beta gamma G protein subunits. *J Physiol* 531: 105-115. DOI: 10.1111/j.1469-7793.2001.0105j.x

274. Viard P, Macrez N, Mironneau C, Mironneau J (2001) Involvement of both G protein alphas and beta gamma subunits in beta-adrenergic stimulation of vascular L-type Ca(2+) channels. *Br J Pharmacol* 132: 669-676. DOI: 10.1038/sj.bjp.0703864

275. Dolphin AC (2009) Calcium channel diversity: multiple roles of calcium channel subunits. *Curr Opin Neurobiol* 19: 237-244. DOI: 10.1016/j.conb.2009.06.006

276. Currie KP (2010) Inhibition of Ca2+ channels and adrenal catecholamine release by G protein coupled receptors. *Cell Mol Neurobiol* 30: 1201-1208. DOI: 10.1007/s10571-010-9596-7

277. Sutherland EW, Rall TW (1958) Fractionation and characterization of a cyclic adenine ribonucleotide formed by tissue particles. *J Biol Chem* 232: 1077-1091.

278. Walsh DA, Perkins JP, Krebs EG (1968) An adenosine 3',5'-monophosphate-dependant protein kinase from rabbit skeletal muscle. *J Biol Chem* 243: 3763-3765.

279. Edelman AM, Blumenthal DK, Krebs EG (1987) Protein serine/threonine kinases. *Annu Rev Biochem* 56: 567-613. DOI: 10.1146/annurev.bi.56.070187.003031

280. Shabb JB (2001) Physiological substrates of cAMP-dependent protein kinase. *Chem Rev* 101: 2381-2411. DOI: 10.1021/cr000236l

281. Bos JL (2003) Epac: a new cAMP target and new avenues in cAMP research. *Nat Rev Mol Cell Biol* 4: 733-738. DOI: 10.1038/nrm1197

282. de Rooij J, Zwartkruis FJ, Verheijen MH, Cool RH, Nijman SM, et al. (1998) Epac is a Rap1 guanine-nucleotide-exchange factor directly activated by cyclic AMP. *Nature* 396: 474-477. DOI: 10.1038/24884

283. Kaupp UB, Seifert R (2002) Cyclic nucleotide-gated ion channels. *Physiol Rev* 82: 769-824.

284. Houslay MD (2010) Underpinning compartmentalised cAMP signalling through targeted cAMP breakdown. *Trends Biochem Sci* 35: 91-100. DOI: 10.1016/j.tibs.2009.09.007

285. Lev S, Moreno H, Martinez R, Canoll P, Peles E, et al. (1995) Protein tyrosine kinase PYK2 involved in Ca(2+)-induced regulation of ion channel and MAP kinase functions. *Nature* 376: 737-745. DOI: 10.1038/376737a0

286. Wang KL, Khan MT, Roufogalis BD (1997) Identification and characterization of a calmodulin-binding domain in Ral-A, a Ras-related GTP-binding protein purified from human erythrocyte membrane. *J Biol Chem* 272: 16002-16009. DOI: 10.1074/jbc.272.25.16002

287. Gainetdinov RR, Premont RT, Bohn LM, Lefkowitz RJ, Caron MG (2004) Desensitization of G protein-coupled receptors and neuronal functions. *Annu Rev Neurosci* 27: 107-144. DOI: 10.1146/annurev.neuro.27.070203.144206

288. Dorn GW, 2nd (2009) GRK mythology: G-protein receptor kinases in cardiovascular disease. *J Mol Med* 87: 455-463. DOI: 10.1007/s00109-009-0450-7

289. Metaye T, Gibelin H, Perdrisot R, Kraimps JL (2005) Pathophysiological roles of G-protein-coupled receptor kinases. *Cell Signal* 17: 917-928. DOI: 10.1016/j.cellsig.2005.01.002

290. Penela P, Murga C, Ribas C, Lafarga V, Mayor F, Jr. (2010) The complex G protein-coupled receptor kinase 2 (GRK2) interactome unveils new physiopathological targets. *Br J Pharmacol* 160: 821-832. DOI: 10.1111/j.1476-5381.2010.00727.x

291. Penela P, Ribas C, Mayor F, Jr. (2003) Mechanisms of regulation of the expression and function of G protein-coupled receptor kinases. *Cell Signal* 15: 973-981. DOI: 10.1016/S0898-6568(03)00099-8

292. Palczewski K, Buczylko J, Lebioda L, Crabb JW, Polans AS (1993) Identification of the N-terminal region in rhodopsin kinase involved in its interaction with rhodopsin. *J Biol Chem* 268: 6004-6013.

293. Penn RB, Pronin AN, Benovic JL (2000) Regulation of G protein-coupled receptor kinases. *Trends Cardiovasc Med* 10: 81-89. DOI: 10.1016/S1050-1738(00)00053-0

294. Pitcher JA, Inglese J, Higgins JB, Arriza JL, Casey PJ, et al. (1992) Role of beta gamma subunits of G proteins in targeting the beta-adrenergic receptor kinase to membrane-bound receptors. *Science* 257: 1264-1267. DOI: 10.1126/science.1325672

295. Inglese J, Koch WJ, Caron MG, Lefkowitz RJ (1992) Isoprenylation in regulation of signal transduction by G-protein-coupled receptor kinases. *Nature* 359: 147-150. DOI: 10.1038/359147a0

296. Eichmann T, Lorenz K, Hoffmann M, Brockmann J, Krasel C, et al. (2003) The amino-terminal domain of G-protein-coupled receptor kinase 2 is a regulatory Gbeta gamma binding site. *J Biol Chem* 278: 8052-8057. DOI: 10.1074/jbc.M204795200

297. DebBurman SK, Ptasienski J, Boetticher E, Lomasney JW, Benovic JL, et al. (1995) Lipid-mediated regulation of G protein-coupled receptor kinases 2 and 3. *J Biol Chem* 270: 5742-5747. DOI: 10.1074/jbc.270.11.5742

298. Kohout TA, Lefkowitz RJ (2003) Regulation of G protein-coupled receptor kinases and arrestins during receptor desensitization. *Mol Pharmacol* 63: 9-18. DOI: 10.1124/mol.63.1.9

299. Perry SJ, Lefkowitz RJ (2002) Arresting developments in heptahelical receptor signaling and regulation. *Trends Cell Biol* 12: 130-138. DOI: 10.1016/S0962-8924(01)02239-5

300. Hirsch JA, Schubert C, Gurevich VV, Sigler PB (1999) The 2.8 A crystal structure of visual arrestin: a model for arrestin's regulation. *Cell* 97: 257-269. DOI: 10.1016/S0092-8674(00)80735-7

301. Han M, Gurevich VV, Vishnivetskiy SA, Sigler PB, Schubert C (2001) Crystal structure of beta-arrestin at 1.9 angstrom: Possible mechanism of receptor binding and membrane translocation. *Structure* 9: 869-880. DOI: 10.1016/S0969-2126(01)00644-X

302. Zhan XZ, Gimenez LE, Gurevich VV, Spiller BW (2011) Crystal Structure of Arrestin-3 Reveals the Basis of the Difference in Receptor Binding Between Two Non-visual Subtypes. *J Mol Biol* 406: 467-478. DOI: 10.1016/j.jmb.2010.12.034

303. Shilton BH, McDowell JH, Smith WC, Hargrave PA (2002) The solution structure and activation of visual arrestin studied by small-angle X-ray scattering. *Eur J Biochem* 269: 3801-3809. DOI: 10.1046/j.1432-1033.2002.03071.x

304. Kovoor A, Celver J, Abdryashitov RI, Chavkin C, Gurevich VV (1999) Targeted construction of phosphorylation-independent beta-arrestin mutants with constitutive activity in cells. *J Biol Chem* 274: 6831-6834. DOI: 10.1074/jbc.274.11.6831

305. Celver J, Vishnivetskiy SA, Chavkin C, Gurevich VV (2002) Conservation of the phosphate-sensitive elements in the arrestin family of proteins. *J Biol Chem* 277: 9043-9048. DOI: 10.1074/jbc.M107400200

306. Lin FT, Krueger KM, Kendall HE, Daaka Y, Fredericks ZL, et al. (1997) Clathrin-mediated endocytosis of the beta-adrenergic receptor is regulated by phosphorylation/dephosphorylation of beta-arrestin1. *J Biol Chem* 272: 31051-31057. DOI: 10.1074/jbc.272.49.31051

307. Lin FT, Chen W, Shenoy S, Cong M, Exum ST, et al. (2002) Phosphorylation of beta-arrestin2 regulates its function in internalization of beta(2)-adrenergic receptors. *Biochemistry* 41: 10692-10699. DOI: 10.1021/bi025705n

308. Shenoy SK, McDonald PH, Kohout TA, Lefkowitz RJ (2001) Regulation of receptor fate by ubiquitination of activated beta 2-adrenergic receptor and beta-arrestin. *Science* 294: 1307-1313. DOI: 10.1126/science.1063866

309. Shenoy SK, Barak LS, Xiao K, Ahn S, Berthouze M, et al. (2007) Ubiquitination of beta-arrestin links seven-transmembrane receptor endocytosis and ERK activation. *J Biol Chem* 282: 29549-29562. DOI: 10.1074/jbc.M700852200

310. Shenoy SK, Lefkowitz RJ (2011) beta-Arrestin-mediated receptor trafficking and signal transduction. *Trends Pharmacol Sci* 32: 521-533. DOI: 10.1016/j.tips.2011.05.002

311. Wyatt D, Malik R, Vesecky AC, Marchese A (2011) Small ubiquitin-like modifier modification of arrestin-3 regulates receptor trafficking. *J Biol Chem* 286: 3884-3893. DOI: 10.1074/jbc.M110.152116

312. Ahn S, Nelson CD, Garrison TR, Miller WE, Lefkowitz RJ (2003) Desensitization, internalization, and signaling functions of beta-arrestins demonstrated by RNA interference. *Proc Natl Acad Sci USA* 100: 1740-1744. DOI: 10.1073/pnas.262789099

313. Kohout TA, Lin FS, Perry SJ, Conner DA, Lefkowitz RJ (2001) beta-Arrestin 1 and 2 differentially regulate heptahelical receptor signaling and trafficking. *Proc Natl Acad Sci USA* 98: 1601-1606.

314. Ferguson SS, Downey WE, 3rd, Colapietro AM, Barak LS, Menard L, et al. (1996) Role of beta-arrestin in mediating agonist-promoted G protein-coupled receptor internalization. *Science* 271: 363-366. DOI: 10.1126/science.271.5247.363

315. Oakley RH, Laporte SA, Holt JA, Barak LS, Caron MG (1999) Association of beta-arrestin with G protein-coupled receptors during clathrin-mediated endocytosis dictates the profile of receptor resensitization. *J Biol Chem* 274: 32248-32257. DOI: 10.1074/jbc.274.45.32248

316. Oakley RH, Laporte SA, Holt JA, Caron MG, Barak LS (2000) Differential affinities of visual arrestin, beta arrestin1, and beta arrestin2 for G protein-coupled receptors delineate two major classes of receptors. *J Biol Chem* 275: 17201-17210. DOI: 10.1074/jbc.M910348199

317. Defea K (2008) Beta-arrestins and heterotrimeric G-proteins: collaborators and competitors in signal transduction. *Br J Pharmacol* 153: S298-309. DOI: 10.1038/sj.bjp.0707508

318. Shenoy SK, Lefkowitz RJ (2003) Multifaceted roles of beta-arrestins in the regulation of seven-membrane-spanning receptor trafficking and signalling. *Biochem J* 375: 503-515. DOI: 10.1042/BJ20031076

319. Ahn S, Shenoy SK, Wei H, Lefkowitz RJ (2004) Differential kinetic and spatial patterns of beta-arrestin and G protein-mediated ERK activation by the angiotensin II receptor. *J Biol Chem* 279: 35518-35525. DOI: 10.1074/jbc.M405878200

320. Sneddon WB, Friedman PA (2007) Beta-arrestin-dependent parathyroid hormone-stimulated extracellular signal-regulated kinase activation and parathyroid hormone type 1 receptor internalization. *Endocrinology* 148: 4073-4079. DOI: 10.1210/en.2007-0343

321. DeFea KA, Vaughn ZD, O'Bryan EM, Nishijima D, Dery O, et al. (2000) The proliferative and antiapoptotic effects of substance P are facilitated by formation of a beta -arrestin-dependent scaffolding complex. *Proc Natl Acad Sci USA* 97: 11086-11091. DOI: 10.1073/pnas.190276697

322. Daaka Y, Luttrell LM, Ahn S, Della Rocca GJ, Ferguson SS, et al. (1998) Essential role for G protein-coupled receptor endocytosis in the activation of mitogen-activated protein kinase. *J Biol Chem* 273: 685-688. DOI: 10.1074/jbc.273.2.685

323. Pierce KL, Maudsley S, Daaka Y, Luttrell LM, Lefkowitz RJ (2000) Role of endocytosis in the activation of the extracellular signal-regulated kinase cascade by sequestering and nonsequestering G protein-coupled receptors. *Proc Natl Acad Sci USA* 97: 1489-1494. DOI: 10.1073/pnas.97.4.1489

324. de Gortazar AR, Alonso V, Alvarez-Arroyo MV, Esbrit P (2006) Transient exposure to PTHrP (107-139) exerts anabolic effects through vascular endothelial growth factor receptor 2 in human osteoblastic cells in vitro. *Calcif Tissue Int* 79: 360-369. DOI: 10.1007/s00223-006-0099-y

325. Terrillon S, Bouvier M (2004) Receptor activity-independent recruitment of betaarrestin2 reveals specific signalling modes. *EMBO* J 23: 3950-3961. DOI: 10.1038/sj.emboj.7600387

326. Galandrin S, Oligny-Longpre G, Bouvier M (2007) The evasive nature of drug efficacy: implications for drug discovery. *Trends Pharmacol Sci* 28: 423-430. DOI: 10.1016/j.tips.2007.06.005

327. Kenakin T (1995) Agonist-receptor efficacy. II. Agonist trafficking of receptor signals. *Trends Pharmacol Sci* 16: 232-238. DOI: 10.1016/S0165-6147(00)89032-X

328. Jarpe MB, Knall C, Mitchell FM, Buhl AM, Duzic E, et al. (1998) [D-Arg1,D-Phe5,D-Trp7,9,Leu11]Substance P acts as a biased agonist toward neuropeptide and chemokine receptors. *J Biol Chem* 273: 3097-3104. DOI: 10.1074/jbc.273.5.3097

329. Bohinc BN, Gesty-Palmer D (2011) beta-arrestin-biased agonism at the parathyroid hormone receptor uncouples bone formation from bone resorption. *Endocr Metab Immune Disord Drug Targets* 11: 112-119. DOI: 10.2174/187153011795564151

330. Rozenfeld R, Devi LA (2007) Receptor heterodimerization leads to a switch in signaling: beta-arrestin2-mediated ERK activation by mu-delta opioid receptor heterodimers. *FASEB J* 21: 2455-2465. DOI: 10.1096/fj.06-7793com

331. Gomes I, Jordan BA, Gupta A, Rios C, Trapaidze N, et al. (2001) G protein coupled receptor dimerization: implications in modulating receptor function. *J Mol Med* 79: 226-242. DOI: 10.1007/s001090100219

332. Satake H, Sakai T (2008) Recent advances and perceptions in studies of heterodimerization between G protein-coupled receptors. *Protein Pept Lett* 15: 300-308. DOI: 10.2174/092986608783744207

333. Gonzalez-Maeso J (2011) GPCR oligomers in pharmacology and signaling. *Mol Brain* 4: 20. DOI: 10.1186/1756-6606-4-20

334. Kaczor AA, Selent J (2011) Oligomerization of G protein-coupled receptors: biochemical and biophysical methods. *Curr Med Chem* 18: 4606-4634. DOI: 10.2174/092986711797379285

335. Tadagaki K, Jockers R, Kamal M (2012) History and biological significance of GPCR heteromerization in the neuroendocrine system. *Neuroendocrinology* 95: 223-231. DOI: 10.1159/000330000

336. Rios CD, Jordan BA, Gomes I, Devi LA (2001) G-protein-coupled receptor dimerization: modulation of receptor function. *Pharmacol Ther* 92: 71-87. DOI: 10.1016/S0163-7258(01)00160-7

337. Rozenfeld R, Devi LA (2010) Receptor heteromerization and drug discovery. *Trends Pharmacol Sci* 31: 124-130. DOI: 10.1016/j.tips.2009.11.008

338. Rozenfeld R, Devi LA (2011) Exploring a role for heteromerization in GPCR signalling specificity. *Biochem J* 433: 11-18. DOI: 10.1042/BJ20100458

339. Gomes I, Ijzerman AP, Ye K, Maillet EL, Devi LA (2011) G protein-coupled receptor heteromerization: a role in allosteric modulation of ligand binding. *Mol Pharmacol* 79: 1044-1052. DOI: 10.1124/mol.110.070847

340. Gomes I, Gupta A, Filipovska J, Szeto HH, Pintar JE, et al. (2004) A role for heterodimerization of mu and delta opiate receptors in enhancing morphine analgesia. *Proc Natl Acad Sci USA* 101: 5135-5139. DOI: 10.1073/pnas.0307601101

341. Gomes I, Jordan BA, Gupta A, Trapaidze N, Nagy V, et al. (2000) Heterodimerization of mu and delta opioid receptors: A role in opiate synergy. *J Neurosci* 20: RC110.

342. Hasbi A, Nguyen T, Fan T, Cheng R, Rashid A, et al. (2007) Trafficking of preassembled opioid mu-delta heterooligomer-Gz signaling complexes to the plasma membrane: coregulation by agonists. *Biochemistry* 46: 12997-13009. DOI: 10.1021/bi701436w

343. Breit A, Gagnidze K, Devi LA, Lagace M, Bouvier M (2006) Simultaneous activation of the delta opioid receptor (deltaOR)/sensory neuron-specific receptor-4 (SNSR-4) hetero-oligomer by the mixed bivalent agonist bovine adrenal medulla peptide 22 activates SNSR-4 but inhibits deltaOR signaling. *Mol Pharmacol* 70: 686-696. DOI: 10.1124/mol.106.022897

344. Pfeiffer M, Kirscht S, Stumm R, Koch T, Wu D, et al. (2003) Heterodimerization of substance P and mu-opioid receptors regulates receptor trafficking and resensitization. *J Biol Chem* 278: 51630-51637. DOI: 10.1074/jbc.M307095200

345. Gehlert DR, Schober DA, Morin M, Berglund MM (2007) Co-expression of neuropeptide Y Y1 and Y5 receptors results in heterodimerization and altered functional properties. *Biochem Pharmacol* 74: 1652-1664. DOI: 10.1016/j.bcp.2007.08.017

346. Murat B, Devost D, Andres M, Mion J, Boulay V, et al. (2012) V1b and CRHR1 receptor heterodimerization mediates synergistic biological actions of vasopressin and CRH. *Mol Endocrinol* 26: 502-520. DOI: 10.1210/me.2011-1202

347. Gelman JS, Fricker LD (2010) Hemopressin and Other Bioactive Peptides from Cytosolic Proteins: Are These Non-Classical Neuropeptides? *AAPS J* 12: 279-289. DOI: 10.1208/s12248-010-9186-0

348. Eipper BA, Mains RE, Herbert E (1986) Peptides in the nervous system. *Trends Neurosci* 9: 463-468. DOI: 10.1016/0166-2236(86)90149-9

349. Akil H, Watson SJ, Young E, Lewis ME, Khachaturian H, et al. (1984) Endogenous opioids: biology and function. *Annu Rev Neurosci* 7: 223-255. DOI: 10.1146/annurev.ne.07.030184.001255

350. Breslin MB, Lindberg I, Benjannet S, Mathis JP, Lazure C, et al. (1993) Differential processing of proenkephalin by prohormone convertases 1(3) and 2 and furin. *J Biol Chem* 268: 27084-27093.

351. Fricker LD, Berezniuk I (2011) Endogenous Opioids. In: Pasternak GW, editor. *The Opiate Receptors.* 2 ed. New York: Humana Press. pp. 93-120.

352. Nicolas P, Li CH (1985) Beta-endorphin-(1-27) is a naturally occurring antagonist to etorphine-induced analgesia. *Proc Natl Acad Sci USA* 82: 3178-3181. DOI: 10.1073/pnas.82.10.3178

353. Mains RE, Cullen EI, May V, Eipper BA (1987) The Role of Secretory Granules in Peptide Biosynthesisa. *Ann NY Acad Sci* 493: 278-291. DOI: 10.1111/j.1749-6632.1987.tb27213.x

354. Smith AI, Funder JW (1988) Proopiomelanocortin processing in the pituitary, central nervous system, and peripheral tissues. *Endocr Rev* 9: 159-179. DOI: 10.1210/edrv-9-1-159

355. Rouillé Y, Duguay SJ, Lund K, Furuta M, Gong Q, et al. (1995) Proteolytic Processing Mechanisms in the Biosynthesis of Neuroendocrine Peptides: The Subtilisin-like Proprotein Convertases. *Front Neuroendocrin* 16: 322-361. DOI: 10.1006/frne.1995.1012

356. Zhou A, Webb G, Zhu X, Steiner DF (1999) Proteolytic Processing in the Secretory Pathway. *J Biol Chem* 274: 20745-20748. DOI: 10.1074/jbc.274.30.20745

357. Devi L (1991) Consensus sequence for processing of peptide precursors at monobasic sites. *FEBS Letters* 280: 189-194. DOI: 10.1016/0014-5793(91)80290-J

358. Henrich S, Lindberg I, Bode W, Than ME (2005) Proprotein convertase models based on the crystal structures of furin and kexin: explanation of their specificity. *J Mol Biol* 345: 211-227. DOI: 10.1016/j.jmb.2004.10.050

359. Seidah NG, Prat A (2002) Precursor convertases in the secretory pathway, cytosol and extracellular milieu. *Essays Biochem* 38: 79-94.

360. Arolas JL, Vendrell J, Aviles FX, Fricker LD (2007) Metallocarboxypeptidases: emerging drug targets in biomedicine. *Curr Pharm Des* 13: 349-366. DOI: 10.2174/138161207780162980

361. Morvan J, Tooze SA (2008) Discovery and progress in our understanding of the regulated secretory pathway in neuroendocrine cells. *Histochem Cell Biol* 129: 243-252. DOI: 10.1007/s00418-008-0377-z

362. Urbe S, Dittie AS, Tooze SA (1997) pH-dependent processing of secretogranin II by the endopeptidase PC2 in isolated immature secretory granules. *Biochem J* 321: 65-74.

363. Wu MM, Grabe M, Adams S, Tsien RY, Moore HP, et al. (2001) Mechanisms of pH regulation in the regulated secretory pathway. *J Biol Chem* 276: 33027-33035. DOI: 10.1074/jbc.M103917200

364. Canaff L, Bennett HP, Hendy GN (1999) Peptide hormone precursor processing: getting sorted? *Mol Cell Endocrinol* 156: 1-6. DOI: 10.1016/S0303-7207(99)00129-X

365. Song L, Fricker LD (1995) Purification and characterization of carboxypeptidase D, a novel carboxypeptidase E-like enzyme, from bovine pituitary. *J Biol Chem* 270: 25007-25013. DOI: 10.1074/jbc.270.42.25007

366. Varlamov O, Fricker LD (1998) Intracellular trafficking of metallocarboxypeptidase D in AtT-20 cells: localization to the trans-Golgi network and recycling from the cell surface. *J Cell Sci* 111 (Pt 7): 877-885.

367. Greene D, Das B, Fricker LD (1992) Regulation of carboxypeptidase E. Effect of pH, temperature and Co2+ on kinetic parameters of substrate hydrolysis. *Biochem J* 285 (Pt 2): 613-618.

368. Charles W W (2006) Roles of acetylation and other post-translational modifications in melanocortin function and interactions with endorphins. *Peptides* 27: 453-471. DOI: 10.1016/j.peptides.2005.05.029

369. Eipper BA, Stoffers DA, Mains RE (1992) The biosynthesis of neuropeptides: peptide alpha-amidation. *Ann Rev Neurosci* 15: 57-85. DOI: 10.1146/annurev.ne.15.030192.000421

370. Sossin WS, Fisher JM, Scheller RH (1989) Cellular and molecular biology of neuropeptide processing and packaging. *Neuron* 2: 1407-1417. DOI: 10.1016/0896-6273(89)90186-4

371. Yasothornsrikul S, Greenbaum D, Medzihradszky KF, Toneff T, Bundey R, et al. (2003) Cathepsin L in secretory vesicles functions as a prohormone-processing enzyme for production of the enkephalin peptide neurotransmitter. *Proc Natl Acad Sci USA* 100: 9590-9595. DOI: 10.1073/pnas.1531542100

372. Funkelstein L, Toneff T, Mosier C, Hwang SR, Beuschlein F, et al. (2008) Major role of cathepsin L for producing the peptide hormones ACTH, beta-endorphin, and alpha-MSH, illustrated by protease gene knockout and expression. *J Biol Chem* 283: 35652-35659. DOI: 10.1074/jbc.M709010200

373. Funkelstein L, Toneff T, Hwang SR, Reinheckel T, Peters C, et al. (2008) Cathepsin L participates in the production of neuropeptide Y in secretory vesicles, demonstrated by protease gene knockout and expression. *J Neurochem* 106: 384-391. DOI: 10.1111/j.1471-4159.2008.05408.x

374. Hook V, Yasothornsrikul S, Greenbaum D, Medzihradszky KF, Troutner K, et al. (2004) Cathepsin L and Arg/Lys aminopeptidase: a distinct prohormone processing pathway for the biosynthesis of peptide neurotransmitters and hormones. *Biol Chem* 385: 473-480. DOI: 10.1515/BC.2004.055

375. Pritchard LE, White A (2007) Neuropeptide processing and its impact on melanocortin pathways. *Endocrinology* 148: 4201-4207. DOI: 10.1210/en.2006-1686

376. Pittius CW, Seizinger BR, Mehraein P, Pasi A, Herz A (1983) Proenkephalin-A-derived peptides are present in human brain. *Life Sci* 33: 41-44. DOI: 10.1016/0024-3205(83)90439-3

377. Kilpatrick DL, Howells RD, Noe M, Bailey LC, Udenfriend S (1985) Expression of preproenkephalin-like mRNA and its peptide products in mammalian testis and ovary. *Proc Natl Acad Sci USA* 82: 7467-7469. DOI: 10.1073/pnas.82.21.7467

378. Matsuo H, Miyata A, Mizuno K (1983) Novel C-terminally amidated opioid peptide in human phaeochromocytoma tumour. *Nature* 305: 721-723. DOI: 10.1038/305721a0

379. Seizinger BR, Liebisch DC, Gramsch C, Herz A, Weber E, et al. (1985) Isolation and structure of a novel C-terminally amidated opioid peptide, amidorphin, from bovine adrenal medulla. *Nature* 313: 57-59. DOI: 10.1038/313057a0

380. Hurlbut DE, Evans CJ, Barchas JD, Leslie FM (1987) Pharmacological properties of a proenkephalin A-derived opioid peptide: BAM 18. *Eur J Pharmacol* 138: 359-366. DOI: 10.1016/0014-2999(87)90474-2

381. Lewis RV, Stern AS (1983) Biosynthesis of the enkephalins and enkephalin-containing polypeptides. *Annu Rev Pharmacol Toxicol* 23: 353-372. DOI: 10.1146/annurev.pa.23.040183.002033

382. Hook V, Funkelstein L, Lu D, Bark S, Wegrzyn J, et al. (2008) Proteases for processing proneuropeptides into peptide neurotransmitters and hormones. *Annu Rev Pharmacol Toxicol* 48: 393-423. DOI: 10.1146/annurev.pharmtox.48.113006.094812

383. Perello M, Stuart R, Nillni EA (2008) Prothyrotropin-releasing Hormone Targets Its Processing Products to Different Vesicles of the Secretory Pathway. *J Biol Chem* 283: 19936-19947. DOI: 10.1074/jbc.M800732200

384. Fisher JM, Sossin W, Newcomb R, Scheller RH (1988) Multiple neuropeptides derived from a common precursor are differentially packaged and transported. *Cell* 54: 813-822. DOI: 10.1016/S0092-8674(88)91131-2

385. Leibel RL, Chung WK, Chua SC, Jr. (1997) The molecular genetics of rodent single gene obesities. *J Biol Chem* 272: 31937-31940. DOI: 10.1074/jbc.272.51.31937

386. Lou H, Kim SK, Zaitsev E, Snell CR, Lu B, et al. (2005) Sorting and activity-dependent secretion of BDNF require interaction of a specific motif with the sorting receptor carboxypeptidase e. *Neuron* 45: 245-255. DOI: 10.1016/j.neuron.2004.12.037

387. Hosaka M, Suda M, Sakai Y, Izumi T, Watanabe T, et al. (2004) Secretogranin III binds to cholesterol in the secretory granule membrane as an adapter for chromogranin A. *J Biol Chem* 279: 3627-3634. DOI: 10.1074/jbc.M310104200

388. Tooze SA, Martens GJ, Huttner WB (2001) Secretory granule biogenesis: rafting to the SNARE. *Trends Cell Biol* 11: 116-122. DOI: 10.1016/S0962-8924(00)01907-3

389. Taupenot L, Harper KL, O'Connor DT (2003) The chromogranin-secretogranin family. *N Engl J Med* 348: 1134-1149. DOI: 10.1056/NEJMra021405

390. Fortenberry Y, Liu J, Lindberg I (1999) The role of the 7B2 CT peptide in the inhibition of prohormone convertase 2 in endocrine cell lines. *J Neurochem* 73: 994-1003. DOI: 10.1046/j.1471-4159.1999.0730994.x

391. Gulbenkian S, Merighi A, Wharton J, Varndell IM, Polak JM (1986) Ultrastructural evidence for the coexistence of calcitonin gene-related peptide and substance P in secretory vesicles of peripheral nerves in the guinea pig. *J Neurocytol* 15: 535-542. DOI: 10.1007/BF01611735

392. Zhang X, Nicholas AP, Hokfelt T (1993) Ultrastructural studies on peptides in the dorsal horn of the spinal cord--I. Co-existence of galanin with other peptides in primary afferents in normal rats. *Neuroscience* 57: 365-384. DOI: 10.1016/0306-4522(93)90069-R

393. Merighi A (2002) Costorage and coexistence of neuropeptides in the mammalian CNS. P*rog Neurobiol* 66: 161-190. DOI: 10.1016/S0301-0082(01)00031-4

394. Salio C, Averill S, Priestley JV, Merighi A (2007) Costorage of BDNF and neuropeptides within individual dense-core vesicles in central and peripheral neurons. *Dev Neurobiol* 67: 326-338. DOI: 10.1002/dneu.20358

395. Green R, Shields D (1984) Somatostatin discriminates between the intracellular pathways of secretory and membrane proteins. *J Cell Biol* 99: 97-104. DOI: 10.1083/jcb.99.1.97

396. Kelly RB, Buckley KM, Burgess TL, Carlson SS, Caroni P, et al. (1983) Membrane traffic in neurons and peptide-secreting cells. *Cold Spring Harb Symp Quant Biol* 48: 697-705. DOI: 10.1101/SQB.1983.048.01.073

397. Ludwig M, Leng G (2006) Dendritic peptide release and peptide-dependent behaviours. *Nat Rev Neurosci* 7: 126-136. DOI: 10.1038/nrn1845

398. Xia X, Lessmann V, Martin TF (2009) Imaging of evoked dense-core-vesicle exocytosis in hippocampal neurons reveals long latencies and kiss-and-run fusion events. *J Cell Sci* 122: 75-82. DOI: 10.1242/jcs.034603

399. Eaton BA, Haugwitz M, Lau D, Moore H-PH (2000) Biogenesis of Regulated Exocytotic Carriers in Neuroendocrine Cells. *J Neurosci* 20: 7334-7344.

400. Iversen LL, Lee, C. M., Gilbert, R.F., Hunt, S., Emson P.S. (1980) Regulation of neuropeptide release. *P RoySoc Lond B Bio* 210: 91-111

401. Pritchard L, Turnbull A, White A (2002) Pro-opiomelanocortin processing in the hypothalamus: impact on melanocortin signalling and obesity. *J Endocrinol* 172: 411-421. DOI: 10.1677/joe.0.1720411

402. Bicknell AB (2008) The Tissue-Specific Processing of Pro-Opiomelanocortin. *Journal of Neuroendocrinology* 20: 692-699. DOI: 10.1111/j.1365-2826.2008.01709.x

403. Csuhai E, Little SS, Hersh LB (1995) Inactivation of neuropeptides. *Prog Brain Res* 104: 131-142. DOI: 10.1016/S0079-6123(08)61788-4

404. Johnson GD, Stevenson T, Ahn K (1999) Hydrolysis of peptide hormones by endothelin-converting enzyme-1. A comparison with neprilysin. *J Biol Chem* 274: 4053-4058. DOI: 10.1074/jbc.274.7.4053

405. Padilla BE, Cottrell GS, Roosterman D, Pikios S, Muller L, et al. (2007) Endothelin-converting enzyme-1 regulates endosomal sorting of calcitonin receptor-like receptor and beta-arrestins. *J Cell Biol* 179: 981-997. DOI: 10.1083/jcb.200704053

406. Cadena DL, Gill GN (1992) Receptor tyrosine kinases. *FASEB J* 6: 2332-2337.

407. Pleuvry BJ (1991) Opioid receptors and their ligands: natural and unnatural. *Br J Anaesth* 66: 370-380. DOI: 10.1093/bja/66.3.370

408. Hegadoren KM, O'Donnell T, Lanius R, Coupland NJ, Lacaze-Masmonteil N (2009) The role of beta-endorphin in the pathophysiology of major depression. *Neuropeptides* 43: 341-353. DOI: 10.1016/j.npep.2009.06.004

409. Dupont A, Barden N, Cusan L, Merand Y, Labrie F, et al. (1980) beta-Endorphin and met-enkephalins: their distribution, modulation by estrogens and haloperidol, and role in neuroendocrine control. *Fed Proc* 39: 2544-2550.

410. Sprouse-Blum AS, Smith G, Sugai D, Parsa FD (2010) Understanding endorphins and their importance in pain management. *Hawaii Med J* 69: 70-71.

411. Hartwig AC (1991) Peripheral beta-endorphin and pain modulation. *Anesth Prog* 38: 75-78.

412. Hargreaves KM, Dionne RA, Mueller GP (1983) Plasma beta-endorphin-like immunoreactivity, pain and anxiety following administration of placebo in oral surgery patients. *J Dent Res* 62: 1170-1173. DOI: 10.1177/00220345830620111601

413. Hayward MD, Pintar JE, Low MJ (2002) Selective reward deficit in mice lacking beta-endorphin and enkephalin. *J Neurosci* 22: 8251-8258.

414. Rubinstein M, Mogil JS, Japon M, Chan EC, Allen RG, et al. (1996) Absence of opioid stress-induced analgesia in mice lacking beta-endorphin by site-directed mutagenesis. *Proc Natl Acad Sci USA* 93: 3995-4000. DOI: 10.1073/pnas.93.9.3995

415. Nguyen AT, Marquez P, Hamid A, Kieffer B, Friedman TC, et al. (2012) The rewarding action of acute cocaine is reduced in beta-endorphin deficient but not in mu opioid receptor knockout mice. *Eur J Pharmacol* 686: 50-54. DOI: 10.1016/j.ejphar.2012.04.040

416. Lukiw WJ (2006) Endogenous signaling complexity in neuropeptides- leucine- and methionine-enkephalin. *Cell Mol Neurobiol* 26: 1003-1010. DOI: 10.1007/s10571-006-9100-6

417. Kieffer BL, Gaveriaux-Ruff C (2002) Exploring the opioid system by gene knockout. *Prog Neurobiol* 66: 285-306. DOI: 10.1016/S0301-0082(02)00008-4

418. Raynor K, Kong H, Chen Y, Yasuda K, Yu L, et al. (1994) Pharmacological characterization of the cloned kappa-, delta-, and mu-opioid receptors. *Mol Pharmacol* 45: 330-334.

419. Merg F, Filliol D, Usynin I, Bazov I, Bark N, et al. (2006) Big dynorphin as a putative endogenous ligand for the kappa-opioid receptor. *J Neurochem* 97: 292-301. DOI: 10.1111/j.1471-4159.2006.03732.x

420. Tejeda HA, Shippenberg TS, Henriksson R (2012) The dynorphin/kappa-opioid receptor system and its role in psychiatric disorders. *Cell Mol Life Sci* 69: 857-896. DOI: 10.1007/s00018-011-0844-x

421. Goldstein A, Ghazarossian VE (1980) Immunoreactive dynorphin in pituitary and brain. *Proc Natl Acad Sci USA* 77: 6207-6210. DOI: 10.1073/pnas.77.10.6207

422. Wee S, Koob GF (2010) The role of the dynorphin-kappa opioid system in the reinforcing effects of drugs of abuse. *Psychopharmacology* 210: 121-135. DOI: 10.1007/s00213-010-1825-8

423. Shirayama Y, Ishida H, Iwata M, Hazama G-i, Kawahara R, et al. (2004) Stress increases dynorphin immunoreactivity in limbic brain regions and dynorphin antagonism produces antidepressant-like effects. *J Neurochem* 90: 1258-1268. DOI: 10.1111/j.1471-4159.2004.02589.x

424. Xin L, Geller EB, Adler MW (1997) Body Temperature and Analgesic Effects of Selective Muand Kappa Opioid Receptor Agonists Microdialyzed into Rat Brain. *J Pharmacol Exp Ther* 281: 499-507.

425. Nestler EJ, Aghajanian GK (1997) Molecular and cellular basis of addiction. *Science* 278: 58-63. DOI: 10.1126/science.278.5335.58

426. Przewlocki R, Lason W, Konecka AM, Gramsch C, Herz A, et al. (1983) The opioid peptide dynorphin, circadian rhythms, and starvation. *Science* 219: 71-73. DOI: 10.1126/science.6129699

427. Han JS, Xie CW (1984) Dynorphin: potent analgesic effect in spinal cord of the rat. *Sci Sin B* 27: 169-177.

428. Wang Z, Gardell LR, Ossipov MH, Vanderah TW, Brennan MB, et al. (2001) Pronociceptive actions of dynorphin maintain chronic neuropathic pain. *J Neurosci* 21: 1779-1786.

429. Bilkei-Gorzo A, Erk S, Schurmann B, Mauer D, Michel K, et al. (2012) Dynorphins regulate fear memory: from mice to men. *J Neurosci* 32: 9335-9343. DOI: 10.1523/JNEUROSCI.1034-12.2012

430. Goldstein A, Tachibana S, Lowney LI, Hunkapiller M, Hood L (1979) Dynorphin-(1-13), an extraordinarily potent opioid peptide. *Proc Natl Acad Sci USA* 76: 6666-6670. DOI: 10.1073/pnas.76.12.6666

431. Sim LJ, Selley DE, Dworkin SI, Childers SR (1996) Effects of chronic morphine administration on mu opioid receptor-stimulated [35S]GTPgammaS autoradiography in rat brain. *J Neurosci* 16: 2684-2692.

432. Chen Y, Mestek A, Liu J, Hurley JA, Yu L (1993) Molecular cloning and functional expression of a mu-opioid receptor from rat brain. *Mol Pharmacol* 44: 8-12.

433. Gavril W P (2001) Insights into mu opioid pharmacology: The role of mu opioid receptor subtypes. *Life Sci* 68: 2213-2219. DOI: 10.1016/S0024-3205(01)01008-6

434. Al-Hasani R, Bruchas MR (2011) Molecular Mechanisms of Opioid Receptor-dependent Signaling and Behavior. *Anesthesiology* 115: 1363-1381.

435. George SR, Zastawny RL, Briones-Urbina R, Cheng R, Nguyen T, et al. (1994) Distinct distributions of mu, delta and kappa opioid receptor mRNA in rat brain. *Biochem Biophys Res Commun* 205: 1438-1444. DOI: 10.1006/bbrc.1994.2826

436. Law P-Y, Loh HH (1999) Regulation of Opioid Receptor Activities. *J Pharmacol Exp Ther* 289: 607-624.

437. Gaveriaux-Ruff C, Kieffer BL (2002) Opioid receptor genes inactivated in mice: the highlights. *Neuropeptides* 36: 62-71. DOI: 10.1054/npep.2002.0900

438. Becker A, Grecksch G, Brödemann R, Kraus J, Peters B, et al. (2000) Morphine self-administration in μ-opioid receptor-deficient mice. *N-S Arch Pharmacol* 361: 584-589. DOI: 10.1007/s002100000244

439. Ghozland S, Matthes HW, Simonin F, Filliol D, Kieffer BL, et al. (2002) Motivational effects of cannabinoids are mediated by mu-opioid and kappa-opioid receptors. *J Neurosci* 22: 1146-1154.

440. Filliol D, Ghozland S, Chluba J, Martin M, Matthes HW, et al. (2000) Mice deficient for delta- and mu-opioid receptors exhibit opposing alterations of emotional responses. *Nat Genet* 25: 195-200. DOI: 10.1038/76061

441. Spanagel R, Herz A, Shippenberg TS (1992) Opposing tonically active endogenous opioid systems modulate the mesolimbic dopaminergic pathway. *Proc Natl Acad Sci USA* 89: 2046-2050. DOI: 10.1073/pnas.89.6.2046

442. McLaughlin JP, Land BB, Li S, Pintar JE, Chavkin C (2005) Prior activation of kappa opioid receptors by U50,488 mimics repeated forced swim stress to potentiate cocaine place preference conditioning. *Neuropsychopharmacol* 31: 787-794. DOI: 10.1038/sj.npp.1300860

443. Simonin F, Valverde O, Smadja C, Slowe S, Kitchen I, et al. (1998) Disruption of the [kappa]-opioid receptor gene in mice enhances sensitivity to chemical visceral pain, impairs pharmacological actions of the selective [kappa]-agonist U-50,488H and attenuates morphine withdrawal. *EMBO J* 17: 886-897. DOI: 10.1093/emboj/17.4.886

444. Yang Y, Chen M, Lai Y, Gantz I, Yagmurlu A, et al. (2003) Molecular determination of agouti-related protein binding to human melanocortin-4 receptor. *Mol Pharmacol* 64: 94-103. DOI: 10.1124/mol.64.1.94

445. Gantz I, Fong TM (2003) The melanocortin system. *Am J Physiol Endocrinol Metab* 284: E468-474.

446. D'Agostino G, Diano S (2010) Alpha-melanocyte stimulating hormone: production and degradation. *J Mol Med* 88: 1195-1201. DOI: 10.1007/s00109-010-0651-0

447. Castro MG, Morrison E (1997) Post-translational processing of proopiomelanocortin in the pituitary and in the brain. *Crit Rev Neurobiol* 11: 35-57. DOI: 10.1615/CritRevNeurobiol.v11.i1.30

448. Zhou A, Bloomquist BT, Mains RE (1993) The prohormone convertases PC1 and PC2 mediate distinct endoproteolytic cleavages in a strict temporal order during proopiomelanocortin biosynthetic processing. *J Biol Chem* 268: 1763-1769.

449. Biebermann H, Kuhnen P, Kleinau G, Krude H (2012) The neuroendocrine circuitry controlled by POMC, MSH, and AGRP. *Handb Exp Pharmacol* 209: 47-75. DOI: 10.1007/978-3-642-24716-3_3

450. Rossi M, Kim MS, Morgan DGA, Small CJ, Edwards CMB, et al. (1998) A C-Terminal Fragment of Agouti-Related Protein Increases Feeding and Antagonizes the Effect of Alpha-Melanocyte Stimulating Hormone in Vivo. *Endocrinology* 139: 4428-4431. DOI: 10.1210/en.139.10.4428

451. Graham M, Shutter JR, Sarmiento U, Sarosi I, Stark KL (1997) Overexpression of Agrt leads to obesity in transgenic mice. *Nat Genet* 17: 273-274. DOI: 10.1038/ng1197-273

452. Xu Y, Elmquist JK, Fukuda M (2011) Central nervous control of energy and glucose balance: focus on the central melanocortin system. *Ann N Y Acad Sci* 1243: 1-14. DOI: 10.1111/j.1749-6632.2011.06248.x

453. Bertolini A, Tacchi R, Vergoni AV (2009) Brain effects of melanocortins. *Pharmacol Res* 59: 13-47. DOI: 10.1016/j.phrs.2008.10.005

454. Tatemoto K, Carlquist M, Mutt V (1982) Neuropeptide Y--a novel brain peptide with structural similarities to peptide YY and pancreatic polypeptide. *Nature* 296: 659-660. DOI: 10.1038/296659a0

455. Kimmel JR, Hayden LJ, Pollock HG (1975) Isolation and characterization of a new pancreatic polypeptide hormone. *J Biol Chem* 250: 9369-9376.

456. Gehlert DR (2004) Introduction to the reviews on neuropeptide Y. *Neuropeptides* 38: 135-140. DOI: 10.1016/j.npep.2004.07.002

457. Milgram SL, Chang EY, Mains RE (1996) Processing and routing of a membrane-anchored form of proneuropeptide Y. *Mol Endocrinol* 10: 837-846. DOI: 10.1210/me.10.7.837

458. Chronwall BM, DiMaggio DA, Massari VJ, Pickel VM, Ruggiero DA, et al. (1985) The anatomy of neuropeptide-Y-containing neurons in rat brain. *Neuroscience* 15: 1159-1181. DOI: 10.1016/0306-4522(85)90260-X

459. Kask A, Harro J, von Horsten S, Redrobe JP, Dumont Y, et al. (2002) The neurocircuitry and receptor subtypes mediating anxiolytic-like effects of neuropeptide Y. *Neurosci Biobehav Rev* 26: 259-283. DOI: 10.1016/S0149-7634(01)00066-5

460. Kalra SP, Dube MG, Pu S, Xu B, Horvath TL, et al. (1999) Interacting appetite-regulating pathways in the hypothalamic regulation of body weight. *Endocr Rev* 20: 68-100. DOI: 10.1210/er.20.1.68

461. Brothers SP, Wahlestedt C (2010) Therapeutic potential of neuropeptide Y (NPY) receptor ligands. *EMBO Mol Med* 2: 429-439. DOI: 10.1002/emmm.201000100

462. Hokfelt T, Brumovsky P, Shi T, Pedrazzini T, Villar M (2007) NPY and pain as seen from the histochemical side. *Peptides* 28: 365-372. DOI: 10.1016/j.peptides.2006.07.024

463. Pedrazzini T, Seydoux J, Kunstner P, Aubert JF, Grouzmann E, et al. (1998) Cardiovascular response, feeding behavior and locomotor activity in mice lacking the NPY Y1 receptor. *Nat Med* 4: 722-726. DOI: 10.1038/nm0698-722

464. Clark JT, Kalra PS, Crowley WR, Kalra SP (1984) Neuropeptide Y and human pancreatic polypeptide stimulate feeding behavior in rats. *Endocrinology* 115: 427-429. DOI: 10.1210/endo-115-1-427

465. Stanley BG, Leibowitz SF (1984) Neuropeptide Y: stimulation of feeding and drinking by injection into the paraventricular nucleus. *Life Sci* 35: 2635-2642. DOI: 10.1016/0024-3205(84)90032-8

466. Stanley BG, Leibowitz SF (1985) Neuropeptide Y injected in the paraventricular hypothalamus: a powerful stimulant of feeding behavior. *Proc Natl Acad Sci USA* 82: 3940-3943. DOI: 10.1073/pnas.82.11.3940

467. Zarjevski N, Cusin I, Vettor R, Rohner-Jeanrenaud F, Jeanrenaud B (1993) Chronic intracerebroventricular neuropeptide-Y administration to normal rats mimics hormonal and metabolic changes of obesity. *Endocrinology* 133: 1753-1758. DOI: 10.1210/en.133.4.1753

468. Lin S, Boey D, Herzog H (2004) NPY and Y receptors: lessons from transgenic and knockout models. *Neuropeptides* 38: 189-200. DOI: 10.1016/j.npep.2004.05.005

469. Erickson JC, Clegg KE, Palmiter RD (1996) Sensitivity to leptin and susceptibility to seizures of mice lacking neuropeptide Y. *Nature* 381: 415-421. DOI: 10.1038/381415a0

470. Bouchard P, Dumont Y, Fournier A, St-Pierre S, Quirion R (1993) Evidence for in vivo interactions between neuropeptide Y-related peptides and sigma receptors in the mouse hippocampal formation. *J Neurosci* 13: 3926-3931.

471. Wu G, Feder A, Wegener G, Bailey C, Saxena S, et al. (2011) Central functions of neuropeptide Y in mood and anxiety disorders. *Expert Opin Ther Targets* 15: 1317-1331. DOI: 10.1517/14728222.2011.628314

472. Gelfo F, Tirassa P, De Bartolo P, Croce N, Bernardini S, et al. (2012) NPY intraperitoneal injections produce antidepressant-like effects and downregulate BDNF in the rat hypothalamus. *CNS Neurosci Ther* 18: 487-492. DOI: 10.1111/j.1755-5949.2012.00314.x

473. Yulyaningsih E, Zhang L, Herzog H, Sainsbury A (2011) NPY receptors as potential targets for anti-obesity drug development. *Br J Pharmacol* 163: 1170-1202. DOI: 10.1111/j.1476-5381.2011.01363.x

474. Kalra SP, Dube MG, Sahu A, Phelps CP, Kalra PS (1991) Neuropeptide Y secretion increases in the paraventricular nucleus in association with increased appetite for food. *Proc Natl Acad Sci USA* 88: 10931-10935. DOI: 10.1073/pnas.88.23.10931

475. Gerald C, Walker MW, Criscione L, Gustafson EL, Batzl-Hartmann C, et al. (1996) A receptor subtype involved in neuropeptide-Y-induced food intake. *Nature* 382: 168-171. DOI: 10.1038/382168a0

476. Schaffhauser AO, Stricker-Krongrad A, Brunner L, Cumin F, Gerald C, et al. (1997) Inhibition of food intake by neuropeptide Y Y5 receptor antisense oligodeoxynucleotides. *Diabetes* 46: 1792-1798. DOI: 10.2337/diabetes.46.11.1792

477. Criscione L, Rigollier P, Batzl-Hartmann C, Rueger H, Stricker-Krongrad A, et al. (1998) Food intake in free-feeding and energy-deprived lean rats is mediated by the neuropeptide Y5 receptor. *J Clin Invest* 102: 2136-2145. DOI: 10.1172/JCI4188

478. Thiele TE, Koh MT, Pedrazzini T (2002) Voluntary alcohol consumption is controlled via the neuropeptide Y Y1 receptor. *J Neurosci* 22: RC208.

479. Sainsbury A, Schwarzer C, Couzens M, Fetissov S, Furtinger S, et al. (2002) Important role of hypothalamic Y2 receptors in body weight regulation revealed in conditional knockout mice. *Proc Natl Acad Sci USA* 99: 8938-8943. DOI: 10.1073/pnas.132043299

480. Naveilhan P, Hassani H, Canals JM, Ekstrand AJ, Larefalk A, et al. (1999) Normal feeding behavior, body weight and leptin response require the neuropeptide Y Y2 receptor. *Nat Med* 5: 1188-1193. DOI: 10.1038/13514

481. Baldock PA, Sainsbury A, Couzens M, Enriquez RF, Thomas GP, et al. (2002) Hypothalamic Y2 receptors regulate bone formation. *J Clin Invest* 109: 915-921.

482. Smith-White MA, Herzog H, Potter EK (2002) Cardiac function in neuropeptide Y Y4 receptor-knockout mice. *Regul Pept* 110: 47-54. DOI: 10.1016/S0167-0115(02)00160-X

483. Marsh DJ, Hollopeter G, Kafer KE, Palmiter RD (1998) Role of the Y5 neuropeptide Y receptor in feeding and obesity. *Nat Med* 4: 718-721. DOI: 10.1038/nm0698-718

484. Kanatani A, Mashiko S, Murai N, Sugimoto N, Ito J, et al. (2000) Role of the Y1 receptor in the regulation of neuropeptide Y-mediated feeding: comparison of wild-type, Y1 receptor-deficient, and Y5 receptor-deficient mice. *Endocrinology* 141: 1011-1016. DOI: 10.1210/en.141.3.1011

485. Pantaleo N, Chadwick W, Park SS, Wang L, Zhou Y, et al. (2010) The mammalian tachykinin ligand-receptor system: an emerging target for central neurological disorders. *CNS Neurol Disord Drug Targets* 9: 627-635. DOI: 10.2174/187152710793361504

486. Regoli D, Boudon A, Fauchére JL (1994) Receptors and antagonists for substance P and related peptides. *Pharmacol Rev* 46: 551-599.

487. Harrison S, Geppetti P (2001) Substance P. *Int J Biochem Cell Biol* 33: 555-576. DOI: 10.1016/S1357-2725(01)00031-0

488. Severini C, Improta G, Falconieri-Erspamer G, Salvadori S, Erspamer V (2002) The Tachykinin Peptide Family. *Pharmacol Rev* 54: 285-322. DOI: 10.1124/pr.54.2.285

489. Gamse R, Holzer P, Lembeck F (1980) Decrease of substance P in primary afferent neurones and impairment of neurogenic plasma extravasation by capsaicin. *Br J Pharmacol* 68: 207-213. DOI: 10.1111/j.1476-5381.1980.tb10409.x

490. Felipe CD, Herrero JF, O'Brien JA, Palmer JA, Doyle CA, et al. (1998) Altered nociception, analgesia and aggression in mice lacking the receptor for substance P. *Nature* 392: 394-397. DOI: 10.1038/32904

491. Quartara L, Maggi CA (1997) The tachykinin NK1 receptor. Part I: Ligands and mechanisms of cellular activation. *Neuropeptides* 31: 537-563. DOI: 10.1016/S0143-4179(97)90001-9

492. Nigel M P (2005) New challenges in the study of the mammalian tachykinins. *Peptides* 26: 1356-1368. DOI: 10.1016/j.peptides.2005.03.030

493. Satake H, Kawada T (2006) Overview of the primary structure, tissue-distribution, and functions of tachykinins and their receptors. *Curr Drug Targets* 7: 963-974. DOI: 10.2174/138945006778019273

494. Alois S (1999) The tachykinin NK1 receptor in the brain: pharmacology and putative functions. *Eur J Pharmacol* 375: 51-60. DOI: 10.1016/S0014-2999(99)00259-9

495. Humphrey JM (2003) Medicinal chemistry of selective neurokinin-1 antagonists. *Curr Top Med Chem* 3: 1423-1435. DOI: 10.2174/1568026033451925

496. Cao YQ, Mantyh PW, Carlson EJ, Gillespie A-M, Epstein CJ, et al. (1998) Primary afferent tachykinins are required to experience moderate to intense pain. *Nature* 392: 390-394. DOI: 10.1038/32897

497. Caterina MJ, Schumacher MA, Tominaga M, Rosen TA, Levine JD, et al. (1997) The capsaicin receptor: a heat-activated ion channel in the pain pathway. *Nature* 389: 816-824. DOI: 10.1038/39807

498. Castagliuolo I, Riegler M, Pasha A, Nikulasson S, Lu B, et al. (1998) Neurokinin-1 (NK-1) receptor is required in Clostridium difficile- induced enteritis. *J Clin Invest* 101: 1547-1550. DOI: 10.1172/JCI2039

499. Zimmer A, Zimmer AM, Baffi J, Usdin T, Reynolds K, et al. (1998) Hypoalgesia in mice with a targeted deletion of the tachykinin 1 gene. *Proc Natl Acad Sci USA* 95: 2630-2635. DOI: 10.1073/pnas.95.5.2630

500. Seto S, Tanioka A, Ikeda M, Izawa S (2005) Design and synthesis of novel 9-substituted-7-aryl-3,4,5,6-tetrahydro-2H-pyrido[4,3-b]- and [2,3-b]-1,5-oxazocin-6-ones as NK1 antagonists. *Bioorg Med Chem Lett* 15: 1479-1484. DOI: 10.1016/j.bmcl.2004.12.091

501. Jordan K (2006) Neurokinin-1-Receptor Antagonists: A New Approach in Antiemetic Therapy. *Onkologie* 29: 39-43. DOI: 10.1159/000089800

502. George DT, Gilman J, Hersh J, Thorsell A, Herion D, et al. (2008) Neurokinin 1 Receptor Antagonism as a Possible Therapy for Alcoholism. *Science* 319: 1536-1539. DOI: 10.1126/science.1153813

503. Yu YJ, Arttamangkul S, Evans CJ, Williams JT, von Zastrow M (2009) Neurokinin 1 Receptors Regulate Morphine-Induced Endocytosis and Desensitization of μ-Opioid Receptors in CNS Neurons. *J Neurosci* 29: 222-233. DOI: 10.1523/JNEUROSCI.4315-08.2009

504. Richter D (1988) Molecular events in expression of vasopressin and oxytocin and their cognate receptors. *Am J Physiol* 255: F207-219.

505. Ross HE, Cole CD, Smith Y, Neumann ID, Landgraf R, et al. (2009) Characterization of the oxytocin system regulating affiliative behavior in female prairie voles. *Neuroscience* 162: 892-903. DOI: 10.1016/j.neuroscience.2009.05.055

506. Gimpl G, Fahrenholz F (2001) The Oxytocin Receptor System: Structure, Function, and Regulation. *Physiol Rev* 81: 629-683.

507. Bancroft J (2005) The endocrinology of sexual arousal. *J Endocrinol* 186: 411-427. DOI: 10.1677/joe.1.06233

508. Young LJ, Wang Z, Insel TR (1998) Neuroendocrine bases of monogamy. *Trends Neurosci* 21: 71-75. DOI: 10.1016/S0166-2236(97)01167-3

509. van Leengoed E, Kerker E, Swanson HH (1987) Inhibition of post-partum maternal behaviour in the rat by injecting an oxytocin antagonist into the cerebral ventricles. *J Endocrinol* 112: 275-282. DOI: 10.1677/joe.0.1120275

510. Kendrick KM (2004) The neurobiology of social bonds. *J Neuroendocrinol* 16: 1007-1008. DOI: 10.1111/j.1365-2826.2004.01262.x

511. Insel TR (1997) A neurobiological basis of social attachment. *Am J Psychiatry* 154: 726-735.

512. Lee HJ, Macbeth AH, Pagani JH, Young WS, 3rd (2009) Oxytocin: the great facilitator of life. *Prog Neurobiol* 88: 127-151.

513. Pollak SD, Seltzer LJ (2009) Attachment and neuroendocrine profiles in infant and adult primates. *Behav Brain Sci* 32: 41-42. DOI: 10.1017/S0140525X09000235

514. Chini B, Manning M (2007) Agonist selectivity in the oxytocin/vasopressin receptor family: new insights and challenges. *Biochem Soc Trans* 35: 737-741. DOI: 10.1042/BST0350737

515. Rodrigues SM, Saslow LR, Garcia N, John OP, Keltner D (2009) Oxytocin receptor genetic variation relates to empathy and stress reactivity in humans. *Proc Natl Acad Sci USA* 106: 21437-21441. DOI: 10.1073/pnas.0909579106

516. Campbell P, Ophir AG, Phelps SM (2009) Central vasopressin and oxytocin receptor distributions in two species of singing mice. *J Comp Neurol* 516: 321-333. DOI: 10.1002/cne.22116

517. Nielsen S, Chou CL, Marples D, Christensen EI, Kishore BK, et al. (1995) Vasopressin increases water permeability of kidney collecting duct by inducing translocation of aquaporin-CD water channels to plasma membrane. *Proc Natl Acad Sci USA* 92: 1013-1017. DOI: 10.1073/pnas.92.4.1013

518. Wiltshire T, Maixner W, Diatchenko L (2011) Relax, you won't feel the pain. *Nat Neurosci* 14: 1496-1497. DOI: 10.1038/nn.2987

519. Wersinger SR, Caldwell HK, Martinez L, Gold P, Hu SB, et al. (2007) Vasopressin 1a receptor knockout mice have a subtle olfactory deficit but normal aggression. *Genes Brain Behav* 6: 540-551. DOI: 10.1111/j.1601-183X.2006.00281.x

520. Parker KJ, Lee TM (2001) Central vasopressin administration regulates the onset of facultative paternal behavior in microtus pennsylvanicus (Meadow Voles). *Horm Behav* 39: 285-294. DOI: 10.1006/hbeh.2001.1655

521. Spanakis E, Milord E, Gragnoli C (2008) AVPR2 variants and mutations in nephrogenic diabetes insipidus: Review and missense mutation significance. *J Cell Physiol* 217: 605-617. DOI: 10.1002/jcp.21552

522. Caldwell HK, Lee H-J, Macbeth AH, Young Iii WS (2008) Vasopressin: Behavioral roles of an "original" neuropeptide. *Prog Neurobiol* 84: 1-24. DOI: 10.1016/j.pneurobio.2007.10.007

523. Keverne EB, Curley JP (2004) Vasopressin, oxytocin and social behaviour. *Curr Opin Neurobiol* 14: 777-783. DOI: 10.1016/j.conb.2004.10.006

524. Bielsky IF, Hu SB, Szegda KL, Westphal H, Young LJ (2004) Profound impairment in social recognition and reduction in anxiety-like behavior in vasopressin V1a receptor knockout mice. *Neuropsychopharmacol* 29: 483-493. DOI: 10.1038/sj.npp.1300360

525. Cheng Y, Hitchcock SA (2007) Targeting cannabinoid agonists for inflammatory and neuropathic pain. *Exp Opin Investig Drugs* 16: 951-965. DOI: 10.1517/13543784.16.7.951

526. Pertwee RG (2006) The pharmacology of cannabinoid receptors and their ligands: an overview. *Int J Obes* 30: S13-18. DOI: 10.1038/sj.ijo.0803272

527. Gomes I, Grushko JS, Golebiewska U, Hoogendoorn S, Gupta A, et al. (2009) Novel endogenous peptide agonists of cannabinoid receptors. *FASEB J* 23: 3020-3029. DOI: 10.1096/fj.09-132142

528. Gelman JS, Sironi J, Castro LM, Ferro ES, Fricker LD (2010) Hemopressins and other hemoglobin-derived peptides in mouse brain: Comparison between brain, blood, and heart peptidome and regulation in Cpe(fat/fat) mice. *J Neurochem* 113: 871-880. DOI: 10.1111/j.1471-4159.2010.06653.x

529. He Y, Hua Y, Keep RF, Liu W, Wang MM, et al. (2011) Hemoglobin expression in neurons and glia after intracerebral hemorrhage. *Acta Neurochir* 111: 133-137. DOI: 10.1007/978-3-7091-0693-8_22

530. Richter F, Meurers BH, Zhu C, Medvedeva VP, Chesselet MF (2009) Neurons express hemoglobin alpha- and beta-chains in rat and human brains. *J Comp Neurol* 515: 538-547. DOI: 10.1002/cne.22062

531. Dale CS, Pagano RdL, Rioli V, Hyslop S, Giorgi R, et al. (2005) Antinociceptive action of hemopressin in experimental hyperalgesia. *Peptides* 26: 431-436. DOI: 10.1016/j.peptides.2004.10.026

532. Dodd GT, Mancini G, Lutz B, Luckman SM (2010) The Peptide Hemopressin Acts through CB1 Cannabinoid Receptors to Reduce Food Intake in Rats and Mice. *J Neurosci* 30: 7369-7376. DOI: 10.1523/JNEUROSCI.5455-09.2010

533. Heimann AS, Gomes L, Dale CS, Pagano RL, Gupta A, et al. (2007) Hemopressin is an inverse agonist of CB1 cannabinoid receptors. *Proc Natl Acad Sci USA* 104: 20588-20593. DOI: 10.1073/pnas.0706980105

534. Matsuda LA, Lolait SJ, Brownstein MJ, Young AC, Bonner TI (1990) Structure of a cannabinoid receptor and functional expression of the cloned cDNA. *Nature* 346: 561-564. DOI: 10.1038/346561a0

535. Howlett AC, Bidaut-Russell M, Devane WA, Melvin LS, Johnson MR, et al. (1990) The cannabinoid receptor: biochemical, anatomical and behavioral characterization. *Trends Neurosci* 13: 420-423. DOI: 10.1016/0166-2236(90)90124-S

536. Pacher P, Mechoulam R (2011) Is lipid signaling through cannabinoid 2 receptors part of a protective system? *Prog Lipid Res* 50: 193-211. DOI: 10.1016/j.plipres.2011.01.001

537. Valverde O, Torrens M (2012) Cbi Receptor-Deficient Mice as a Model for Depression. *Neuroscience* 204: 193-206. DOI: 10.1016/j.neuroscience.2011.09.031

538. Zimmer A, Zimmer AM, Hohmann AG, Herkenham M, Bonner TI (1999) Increased mortality, hypoactivity, and hypoalgesia in cannabinoid CB1 receptor knockout mice. *Proc Natl Acad Sci USA* 96: 5780-5785. DOI: 10.1073/pnas.96.10.5780

539. Buckley NE (2008) The peripheral cannabinoid receptor knockout mice: an update. *Br J Pharmacol* 153: 309-318. DOI: 10.1038/sj.bjp.0707527

540. Allen SJ, Watson JJ, Dawbarn D (2011) The neurotrophins and their role in Alzheimer's disease. *Curr Neuropharmacol* 9: 559-573. DOI: 10.2174/157015911798376190

541. Hibbert AP, Morris SJ, Seidah NG, Murphy RA (2003) Neurotrophin-4, alone or heterodimerized with brain-derived neurotrophic factor, is sorted to the constitutive secretory pathway. *J Biol Chem* 278: 48129-48136. DOI: 10.1074/jbc.M300961200

542. Mowla SJ, Farhadi HF, Pareek S, Atwal JK, Morris SJ, et al. (2001) Biosynthesis and post-translational processing of the precursor to brain-derived neurotrophic factor. *J Biol Chem* 276: 12660-12666. DOI: 10.1074/jbc.M008104200

543. Freeman RS, Burch RL, Crowder RJ, Lomb DJ, Schoell MC, et al. (2004) NGF deprivation-induced gene expression: after ten years, where do we stand? *Prog Brain Res* 146: 111-126. DOI: 10.1016/S0079-6123(03)46008-1

544. Madduri S, Papaloizos M, Gander B (2009) Synergistic effect of GDNF and NGF on axonal branching and elongation in vitro. *Neurosci Res* 65: 88-97. DOI: 10.1016/j.neures.2009.06.003

545. Pae CU, Marks DM, Han C, Patkar AA, Steffens D (2008) Does neurotropin-3 have a therapeutic implication in major depression? *Int J Neurosci* 118: 1515-1522. DOI: 10.1080/00207450802174589

546. Xie CW, Sayah D, Chen QS, Wei WZ, Smith D, et al. (2000) Deficient long-term memory and long-lasting long-term potentiation in mice with a targeted deletion of neurotrophin-4 gene. *Proc Natl Acad Sci USA* 97: 8116-8121. DOI: 10.1073/pnas.140204597

547. Tsao D, Thomsen HK, Chou J, Stratton J, Hagen M, et al. (2008) TrkB agonists ameliorate obesity and associated metabolic conditions in mice. *Endocrinology* 149: 1038-1048. DOI: 10.1210/en.2007-1166

548. Acheson A, Conover JC, Fandl JP, DeChiara TM, Russell M, et al. (1995) A BDNF autocrine loop in adult sensory neurons prevents cell death. *Nature* 374: 450-453. DOI: 10.1038/374450a0

549. Huang EJ, Reichardt LF (2001) Neurotrophins: roles in neuronal development and function. *Annu Rev Neurosci* 24: 677-736. DOI: 10.1146/annurev.neuro.24.1.677

550. Bekinschtein P, Cammarota M, Katche C, Slipczuk L, Rossato JI, et al. (2008) BDNF is essential to promote persistence of long-term memory storage. *Proc Natl Acad Sci USA* 105: 2711-2716. DOI: 10.1073/pnas.0711863105

551. Segal RA (2003) Selectivity in neurotrophin signaling: theme and variations. *Annu Rev Neurosci* 26: 299-330. DOI: 10.1146/annurev.neuro.26.041002.131421

552. Russo SJ, Mazei-Robison MS, Ables JL, Nestler EJ (2009) Neurotrophic factors and structural plasticity in addiction. *Neuropharmacology* 56: 73-82. DOI: 10.1016/j.neuropharm.2008.06.059

553. Kaplan DR, Hempstead BL, Martin-Zanca D, Chao MV, Parada LF (1991) The trk proto-oncogene product: a signal transducing receptor for nerve growth factor. *Science* 252: 554-558. DOI: 10.1126/science.1850549

554. Klein R, Jing SQ, Nanduri V, O'Rourke E, Barbacid M (1991) The trk proto-oncogene encodes a receptor for nerve growth factor. *Cell* 65: 189-197. DOI: 10.1016/0092-8674(91)90419-Y

555. Huang EJ, Reichardt LF (2003) Trk receptors: roles in neuronal signal transduction. *Annu Rev Biochem* 72: 609-642. DOI: 10.1146/annurev.biochem.72.121801.161629

556. Rifkin JT, Todd VJ, Anderson LW, Lefcort F (2000) Dynamic expression of neurotrophin receptors during sensory neuron genesis and differentiation. *Dev Biol* 227: 465-480. DOI: 10.1006/dbio.2000.9841

557. Averill S, McMahon SB, Clary DO, Reichardt LF, Priestley JV (1995) Immunocytochemical localization of trkA receptors in chemically identified subgroups of adult rat sensory neurons. *Eur J Neurosci* 7: 1484-1494. DOI: 10.1111/j.1460-9568.1995.tb01143.x

558. Molliver DC, Snider WD (1997) Nerve growth factor receptor TrkA is down-regulated during postnatal development by a subset of dorsal root ganglion neurons. *J Comp Neurol* 381: 428-438. DOI: 10.1002/(SICI)1096-9861(19970519)381:4<428::AID-CNE3>3.0.CO;2-4

559. Berkemeier LR, Winslow JW, Kaplan DR, Nikolics K, Goeddel DV, et al. (1991) Neurotrophin-5: a novel neurotrophic factor that activates trk and trkB. *Neuron* 7: 857-866. DOI: 10.1016/0896-6273(91)90287-A

560. Fryer RH, Kaplan DR, Feinstein SC, Radeke MJ, Grayson DR, et al. (1996) Developmental and mature expression of full-length and truncated TrkB receptors in the rat forebrain. *J Comp Neurol* 374: 21-40. DOI: 10.1002/(SICI)1096-9861(19961007)374:1<21::AID-CNE2>3.0.CO;2-P

561. Shelton DL, Sutherland J, Gripp J, Camerato T, Armanini MP, et al. (1995) Human trks: molecular cloning, tissue distribution, and expression of extracellular domain immunoadhesins. *J Neurosci* 15: 477-491.

562. Chakravarthy S, Saiepour MH, Bence M, Perry S, Hartman R, et al. (2006) Postsynaptic TrkB signaling has distinct roles in spine maintenance in adult visual cortex and hippocampus. *Proc Natl Acad Sci USA* 103: 1071-1076. DOI: 10.1073/pnas.0506305103

563. Danzer SC, Kotloski RJ, Walter C, Hughes M, McNamara JO (2008) Altered morphology of hippocampal dentate granule cell presynaptic and postsynaptic terminals following conditional deletion of TrkB. *Hippocampus* 18: 668-678. DOI: 10.1002/hipo.20426

564. Horch HW, Kruttgen A, Portbury SD, Katz LC (1999) Destabilization of cortical dendrites and spines by BDNF. *Neuron* 23: 353-364. DOI: 10.1016/S0896-6273(00)80785-0

565. von Bohlen und Halbach O, Minichiello L, Unsicker K (2008) TrkB but not trkC receptors are necessary for postnatal maintenance of hippocampal spines. *Neurobiol Aging* 29: 1247-1255. DOI: 10.1016/j.neurobiolaging.2007.02.028

566. Patterson SL, Abel T, Deuel TA, Martin KC, Rose JC, et al. (1996) *Neuron* 16: 1137-1145. DOI: 10.1016/S0896-6273(00)80140-3

567. Zakharenko SS, Patterson SL, Dragatsis I, Zeitlin SO, Siegelbaum SA, et al. (2003) Presynaptic BDNF required for a presynaptic but not postsynaptic component of LTP at hippocampal CA1-CA3 synapses. *Neuron* 39: 975-990. DOI: 10.1016/S0896-6273(03)00543-9

568. Korte M, Kang H, Bonhoeffer T, Schuman E (1998) A role for BDNF in the late-phase of hippocampal long-term potentiation. *Neuropharmacology* 37: 553-559. DOI: 10.1016/S0028-3908(98)00035-5

569. Ernfors P, Kucera J, Lee KF, Loring J, Jaenisch R (1995) Studies on the physiological role of brain-derived neurotrophic factor and neurotrophin-3 in knockout mice. *Int J Dev Biol* 39: 799-807.

570. Lamballe F, Klein R, Barbacid M (1991) trkC, a new member of the trk family of tyrosine protein kinases, is a receptor for neurotrophin-3. *Cell* 66: 967-979. DOI: 10.1016/0092-8674(91)90442-2

571. Wright DE, Snider WD (1995) Neurotrophin receptor mRNA expression defines distinct populations of neurons in rat dorsal root ganglia. *J Comp Neurol* 351: 329-338. DOI: 10.1002/cne.903510302

572. Scherer T, Buettner C (2011) Yin and Yang of hypothalamic insulin and leptin signaling in regulating white adipose tissue metabolism. *Rev Endocr Metab Disord* 12: 235-243. DOI: 10.1007/s11154-011-9190-4

573. Roubos EW, Dahmen M, Kozicz T, Xu L (2012) Leptin and the hypothalamo-pituitary-adrenal stress axis. *Gen Comp Endocrinol* 177: 28-36. DOI: 10.1016/j.ygcen.2012.01.009

574. Elias CF, Purohit D (2013) Leptin signaling and circuits in puberty and fertility. *Cell Mol Life Sci* 70: 841-862. DOI: 10.1007/s00018-012-1095-1

575. Margetic S, Gazzola C, Pegg GG, Hill RA (2002) Leptin: a review of its peripheral actions and interactions. *Int J Obes Relat Metab Disord* 26: 1407-1433. DOI: 10.1038/sj.ijo.0802142

576. Morash B, Li A, Murphy PR, Wilkinson M, Ur E (1999) Leptin gene expression in the brain and pituitary gland. *Endocrinology* 140: 5995-5998. DOI: 10.1210/en.140.12.5995

577. Brennan AM, Mantzoros CS (2006) Drug Insight: the role of leptin in human physiology and pathophysiology--emerging clinical applications. *Nat Clin Pract Endocrinol Metab* 2: 318-327. DOI: 10.1038/ncpendmet0196

578. Buettner C, Muse ED, Cheng A, Chen L, Scherer T, et al. (2008) Leptin controls adipose tissue lipogenesis via central, STAT3-independent mechanisms. *Nat Med* 14: 667-675. DOI: 10.1038/nm1775

579. Banks WA, Kastin AJ, Huang W, Jaspan JB, Maness LM (1996) Leptin enters the brain by a saturable system independent of insulin. *Peptides* 17: 305-311. DOI: 10.1016/0196-9781(96)00025-3

580. Kastin AJ, Pan W, Maness LM, Koletsky RJ, Ernsberger P (1999) Decreased transport of leptin across the blood-brain barrier in rats lacking the short form of the leptin receptor. *Peptides* 20: 1449-1453. DOI: 10.1016/S0196-9781(99)00156-4

581. Feng H, Zheng L, Feng Z, Zhao Y, Zhang N (2012) The role of leptin in obesity and the potential for leptin replacement therapy. *Endocrine* (Epub ahead of print) DOI: 10.1007/s12020-012-9865-y

582. Cottrell EC, Mercer JG (2012) Leptin receptors. *Handb Exp Pharmacol* 209: 3-21. DOI: 10.1007/978-3-642-24716-3_1

583. Gibson WT, Farooqi IS, Moreau M, DePaoli AM, Lawrence E, et al. (2004) Congenital leptin deficiency due to homozygosity for the Delta133G mutation: report of another case and evaluation of response to four years of leptin therapy. *J Clin Endocrinol Metab* 89: 4821-4826. DOI: 10.1210/jc.2004-0376

584. Farooqi IS, Bullmore E, Keogh J, Gillard J, O'Rahilly S, et al. (2007) Leptin regulates striatal regions and human eating behavior. *Science* 317: 1355. DOI: 10.1126/science.1144599

585. Cohen P, Zhao C, Cai X, Montez JM, Rohani SC, et al. (2001) Selective deletion of leptin receptor in neurons leads to obesity. *J Clin Invest* 108: 1113-1121.

586. Bell GI, Pictet RL, Rutter WJ, Cordell B, Tischer E, et al. (1980) Sequence of the human insulin gene. *Nature* 284: 26-32. DOI: 10.1038/284026a0

587. van Belle TL, Coppieters KT, von Herrath MG (2011) Type 1 diabetes: etiology, immunology, and therapeutic strategies. *Physiol Rev* 91: 79-118. DOI: 10.1152/physrev.00003.2010

588. Steiner DF, Oyer PE (1967) The biosynthesis of insulin and a probable precursor of insulin by a human islet cell adenoma. *Proc Natl Acad Sci* USA 57: 473-480. DOI: 10.1073/pnas.57.2.473

589. Howell SL, Bird GS (1989) Biosynthesis and secretion of insulin. *Br Med Bull* 45: 19-36.

590. Cawston EE, Miller LJ (2010) Therapeutic potential for novel drugs targeting the type 1 cholecystokinin receptor. *Br J Pharmacol* 159: 1009-1021. DOI: 10.1111/j.1476-5381.2009.00489.x

591. Lin Y, Sun Z (2010) Current views on type 2 diabetes. *J Endocrinol* 204: 1-11. DOI: 10.1677/JOE-09-0260

592. Duckworth WC, Bennett RG, Hamel FG (1998) Insulin degradation: progress and potential. *Endocr Rev* 19: 608-624. DOI: 10.1210/er.19.5.608

593. Youngren JF (2007) Regulation of insulin receptor function. *Cell Mol Life Sci* 64: 873-891. DOI: 10.1007/s00018-007-6359-9

594. Schulingkamp RJ, Pagano TC, Hung D, Raffa RB (2000) Insulin receptors and insulin action in the brain: review and clinical implications. *Neurosci Biobehav Rev* 24: 855-872. DOI: 10.1016/S0149-7634(00)00040-3

595. Plum L, Schubert M, Bruning JC (2005) The role of insulin receptor signaling in the brain. *Trends Endocrinol Metab* 16: 59-65. DOI: 10.1016/j.tem.2005.01.008

596. Kitamura T, Kahn CR, Accili D (2003) Insulin receptor knockout mice. *Annu Rev Physiol* 65: 313-332. DOI: 10.1146/annurev.physiol.65.092101.142540

597. Gupta A, Mulder J, Gomes I, Rozenfeld R, Bushlin I, et al. (2010) Increased abundance of opioid receptor heteromers after chronic morphine administration. *Sci Signal* 3: ra54. DOI: 10.1126/scisignal.2000807

598. Berg KA, Rowan MP, Gupta A, Sanchez TA, Silva M, et al. (2012) Allosteric Interactions between delta and kappa Opioid Receptors in Peripheral Sensory Neurons. *Mol Pharmacol* 81: 264-272. DOI: 10.1124/mol.111.072702

599. Rozenfeld R, Gupta A, Gagnidze K, Lim MP, Gomes I, et al. (2011) AT1R-CBR heteromerization reveals a new mechanism for the pathogenic properties of angiotensin II. *EMBO J* 30: 2350-2363. DOI: 10.1038/emboj.2011.139

600. Wettschureck N, Offermanns S (2005) Mammalian G proteins and their cell type specific functions. *Physiol Rev* 85: 1159-1204. DOI: 10.1152/physrev.00003.2005

601. Wadhawan S, Dickins B, Nekrutenko A (2008) Wheels within wheels: clues to the evolution of the Gnas and Gnal loci. *Mol Biol Evol* 25: 2745-2757. DOI: 10.1093/molbev/msn229

602. Wilkie TM, Gilbert DJ, Olsen AS, Chen XN, Amatruda TT, et al. (1992) Evolution of the mammalian G protein alpha subunit multigene family. *Nat Genet* 1: 85-91. DOI: 10.1038/ng0592-85

603. Ruiz-Velasco V, Ikeda SR, Puhl HL (2002) Cloning, tissue distribution, and functional expression of the human G protein beta 4-subunit. *Physiol Genomics* 8: 41-50.

604. Morishita R, Ueda H, Kato K, Asano T (1998) Identification of two forms of the gamma subunit of G protein, gamma10 and gamma11, in bovine lung and their tissue distribution in the rat. *FEBS Lett* 428: 85-88. DOI: 10.1016/S0014-5793(98)00498-0

605. Wei F, Qiu CS, Kim SJ, Muglia L, Maas JW, et al. (2002) Genetic elimination of behavioral sensitization in mice lacking calmodulin-stimulated adenylyl cyclases. *Neuron* 36: 713-726. DOI: 10.1016/S0896-6273(02)01019-X

606. Li S, Lee ML, Bruchas MR, Chan GC, Storm DR, et al. (2006) Calmodulin-stimulated adenylyl cyclase gene deletion affects morphine responses. *Mol Pharmacol* 70: 1742-1749. DOI: 10.1124/mol.106.025783

607. Maas JW, Jr., Indacochea RA, Muglia LM, Tran TT, Vogt SK, et al. (2005) Calcium-stimulated adenylyl cyclases modulate ethanol-induced neurodegeneration in the neonatal brain. *J Neurosci* 25: 2376-2385. DOI: 10.1523/JNEUROSCI.4940-04.2005

608. Zachariou V, Liu R, LaPlant Q, Xiao G, Renthal W, et al. (2008) Distinct roles of adenylyl cyclases 1 and 8 in opiate dependence: behavioral, electrophysiological, and molecular studies. *Biol Psychiatry* 63: 1013-1021. DOI: 10.1016/j.biopsych.2007.11.021

609. Visel A, Alvarez-Bolado G, Thaller C, Eichele G (2006) Comprehensive analysis of the expression patterns of the adenylate cyclase gene family in the developing and adult mouse brain. *J Comp Neurol* 496: 684-697. DOI: 10.1002/cne.20953

610. Ludwig MG, Seuwen K (2002) Characterization of the human adenylyl cyclase gene family: cDNA, gene structure, and tissue distribution of the nine isoforms. *J Recept Signal Transduct Res* 22: 79-110. DOI: 10.1081/RRS-120014589

611. Wong ST, Trinh K, Hacker B, Chan GC, Lowe G, et al. (2000) Disruption of the type III adenylyl cyclase gene leads to peripheral and behavioral anosmia in transgenic mice. *Neuron* 27: 487-497. DOI: 10.1016/S0896-6273(00)00060-X

612. Okumura S, Takagi G, Kawabe J, Yang G, Lee MC, et al. (2003) Disruption of type 5 adenylyl cyclase gene preserves cardiac function against pressure overload. *Proc Natl Acad Sci USA* 100: 9986-9990. DOI: 10.1073/pnas.1733772100

613. Lee KW, Hong JH, Choi IY, Che Y, Lee JK, et al. (2002) Impaired D2 dopamine receptor function in mice lacking type 5 adenylyl cyclase. *J Neurosci* 22: 7931-7940.

614. Kim KS, Lee KW, Lee KW, Im JY, Yoo JY, et al. (2006) Adenylyl cyclase type 5 (AC5) is an essential mediator of morphine action. *Proc Natl Acad Sci USA* 103: 3908-3913. DOI: 10.1073/pnas.0508812103

615. Tang T, Gao MH, Lai NC, Firth AL, Takahashi T, et al. (2008) Adenylyl cyclase type 6 deletion decreases left ventricular function via impaired calcium handling. *Circulation* 117: 61-69. DOI: 10.1161/CIRCULATIONAHA.107.730069

616. Zhang M, Moon C, Chan GC, Yang L, Zheng F, et al. (2008) Ca-stimulated type 8 adenylyl cyclase is required for rapid acquisition of novel spatial information and for working/episodic-like memory. *J Neurosci* 28: 4736-4744. DOI: 10.1523/JNEUROSCI.1177-08.2008

617. Kim D, Jun KS, Lee SB, Kang NG, Min DS, et al. (1997) Phospholipase C isozymes selectively couple to specific neurotransmitter receptors. *Nature* 389: 290-293. DOI: 10.1038/38508

618. Jiang H, Kuang Y, Wu Y, Xie W, Simon MI, et al. (1997) Roles of phospholipase C beta2 in chemoattractant-elicited responses. *Proc Natl Acad Sci USA* 94: 7971-7975. DOI: 10.1073/pnas.94.15.7971

619. Li Z, Jiang H, Xie W, Zhang Z, Smrcka AV, et al. (2000) Roles of PLC-beta2 and -beta3 and PI3Kgamma in chemoattractant-mediated signal transduction. *Science* 287: 1046-1049. DOI: 10.1126/science.287.5455.1046

620. Zhang Y, Hoon MA, Chandrashekar J, Mueller KL, Cook B, et al. (2003) Coding of sweet, bitter, and umami tastes: different receptor cells sharing similar signaling pathways. *Cell* 112: 293-301. DOI: 10.1016/S0092-8674(03)00071-0

621. Xie W, Samoriski GM, McLaughlin JP, Romoser VA, Smrcka A, et al. (1999) Genetic alteration of phospholipase C beta3 expression modulates behavioral and cellular responses to mu opioids. *Proc Natl Acad Sci USA* 96: 10385-10390. DOI: 10.1073/pnas.96.18.10385

622. Xiao W, Hong H, Kawakami Y, Kato Y, Wu D, et al. (2009) Tumor suppression by phospholipase C-beta3 via SHP-1-mediated dephosphorylation of Stat5. *Cancer Cell* 16: 161-171. DOI: 10.1016/j.ccr.2009.05.018

623. Wang Z, Liu B, Wang P, Dong X, Fernandez-Hernando C, et al. (2008) Phospholipase C beta3 deficiency leads to macrophage hypersensitivity to apoptotic induction and reduction of atherosclerosis in mice. *J Clin Invest* 118: 195-204. DOI: 10.1172/JCI33139

624. Jiang H, Lyubarsky A, Dodd R, Vardi N, Pugh E, et al. (1996) Phospholipase C beta 4 is involved in modulating the visual response in mice. *Proc Natl Acad Sci USA* 93: 14598-14601. DOI: 10.1073/pnas.93.25.14598

625. Hirono M, Sugiyama T, Kishimoto Y, Sakai I, Miyazawa T, et al. (2001) Phospholipase Cbeta4 and protein kinase Calpha and/or protein kinase CbetaI are involved in the induction of long term depression in cerebellar Purkinje cells. *J Biol Chem*m 276: 45236-45242. DOI: 10.1074/jbc.M105413200

626. Nakamura Y, Fukami K, Yu H, Takenaka K, Kataoka Y, et al. (2003) Phospholipase Cdelta1 is required for skin stem cell lineage commitment. *EMBO J* 22: 2981-2991. DOI: 10.1093/emboj/cdg302

627. Nakamura Y, Ichinohe M, Hirata M, Matsuura H, Fujiwara T, et al. (2008) Phospholipase C-delta1 is an essential molecule downstream of Foxn1, the gene responsible for the nude mutation, in normal hair development. *FASEB J* 22: 841-849. DOI: 10.1096/fj.07-9239com

628. Fukami K, Nakao K, Inoue T, Kataoka Y, Kurokawa M, et al. (2001) Requirement of phospholipase Cdelta4 for the zona pellucida-induced acrosome reaction. *Science* 292: 920-923. DOI: 10.1126/science.1059042

629. Fukami K, Yoshida M, Inoue T, Kurokawa M, Fissore RA, et al. (2003) Phospholipase Cdelta4 is required for Ca2+ mobilization essential for acrosome reaction in sperm. *J Cell Biol* 161: 79-88. DOI: 10.1083/jcb.200210057

630. Bai Y, Edamatsu H, Maeda S, Saito H, Suzuki N, et al. (2004) Crucial role of phospholipase Cepsilon in chemical carcinogen-induced skin tumor development. *Cancer Res* 64: 8808-8810. DOI: 10.1158/0008-5472.CAN-04-3143

631. Tadano M, Edamatsu H, Minamisawa S, Yokoyama U, Ishikawa Y, et al. (2005) Congenital semilunar valvulogenesis defect in mice deficient in phospholipase C epsilon. *Mol Cell Biol* 25: 2191-2199. DOI: 10.1128/MCB.25.6.2191-2199.2005

632. Wang H, Oestreich EA, Maekawa N, Bullard TA, Vikstrom KL, et al. (2005) Phospholipase C epsilon modulates beta-adrenergic receptor-dependent cardiac contraction and inhibits cardiac hypertrophy. *Circ Res* 97: 1305-1313. DOI: 10.1161/01.RES.0000196578.15385.bb

633. Kanemaru K, Nakahara M, Nakamura Y, Hashiguchi Y, Kouchi Z, et al. (2010) Phospholipase C-eta2 is highly expressed in the habenula and retina. *Gene Expr Patterns* 10: 119-126. DOI: 10.1016/j.gep.2009.12.004

634. Catterall WA (2011) Voltage-gated calcium channels. *Cold Spring Harb Perspect Biol* 3: a003947. DOI: 10.1101/cshperspect.a003947

635. Dolphin AC (2003) G protein modulation of voltage-gated calcium channels. *Pharmacol Rev* 55: 607-627. DOI: 10.1124/pr.55.4.3

636. Gainetdinov RR, Premont RT, Caron MG, Lefkowitz RJ (2000) Reply: receptor specificity of G-protein-coupled receptor kinases. *Trends Pharmacol Sci* 21: 366-367. DOI: 10.1016/S0165-6147(00)01538-8

637. Loudon RP, Perussia B, Benovic JL (1996) Differentially regulated expression of the G-protein-coupled receptor kinases, betaARK and GRK6, during myelomonocytic cell development in vitro. *Blood* 88: 4547-4557.

638. Lombardi MS, Kavelaars A, Schedlowski M, Bijlsma JW, Okihara KL, et al. (1999) Decreased expression and activity of G-protein-coupled receptor kinases in peripheral blood mononuclear cells of patients with rheumatoid arthritis. *FASEB J* 13: 715-725.

639. Freeman JL, Pitcher JA, Li X, Bennett V, Lefkowitz RJ (2000) alpha-Actinin is a potent regulator of G protein-coupled receptor kinase activity and substrate specificity in vitro. *FEBS Lett* 473: 280-284. DOI: 10.1016/S0014-5793(00)01543-X

640. Pronin AN, Benovic JL (1997) Regulation of the G protein-coupled receptor kinase GRK5 by protein kinase C. *J Biol Chem* 272: 3806-3812. DOI: 10.1074/jbc.272.29.18273

641. Levay K, Satpaev DK, Pronin AN, Benovic JL, Slepak VZ (1998) Localization of the sites for Ca2+-binding proteins on G protein-coupled receptor kinases. *Biochemistry* 37: 13650-13659. DOI: 10.1021/bi980998z

642. Carman CV, Parent JL, Day PW, Pronin AN, Sternweis PM, et al. (1999) Selective regulation of Galpha(q/11) by an RGS domain in the G protein-coupled receptor kinase, GRK2. *J Biol Chem* 274: 34483-34492. DOI: 10.1074/jbc.274.48.34483

643. Sallese M, Mariggio S, D'Urbano E, Iacovelli L, De Blasi A (2000) Selective regulation of Gq signaling by G protein-coupled receptor kinase 2: direct interaction of kinase N terminus with activated galphaq. *Mol Pharmacol* 57: 826-831.

644. Pitcher JA, Freedman NJ, Lefkowitz RJ (1998) G protein-coupled receptor kinases. *Annu Rev Biochem* 67: 653-692. DOI: 10.1146/annurev.biochem.67.1.653

645. Lohse MJ, Krasel C, Winstel R, Mayor F, Jr. (1996) G-protein-coupled receptor kinases. *Kidney Int* 49: 1047-1052. DOI: 10.1038/ki.1996.153

646. Krasel C, Dammeier S, Winstel R, Brockmann J, Mischak H, et al. (2001) Phosphorylation of GRK2 by protein kinase C abolishes its inhibition by calmodulin. *J Biol Chem* 276: 1911-1915. DOI: 10.1074/jbc.M008773200

647. Pitcher JA, Tesmer JJ, Freeman JL, Capel WD, Stone WC, et al. (1999) Feedback inhibition of G protein-coupled receptor kinase 2 (GRK2) activity by extracellular signal-regulated kinases. *J Biol Chem* 274: 34531-34534. DOI: 10.1074/jbc.274.49.34531

648. Fan G, Shumay E, Malbon CC, Wang H (2001) c-Src tyrosine kinase binds the beta 2-adrenergic receptor via phospho-Tyr-350, phosphorylates G-protein-linked receptor kinase 2, and mediates agonist-induced receptor desensitization. *J Biol Chem* 276: 13240-13247. DOI: 10.1074/jbc.M011578200

649. Penela P, Elorza A, Sarnago S, Mayor F, Jr. (2001) Beta-arrestin- and c-Src-dependent degradation of G-protein-coupled receptor kinase 2. *EMBO J* 20: 5129-5138. DOI: 10.1093/emboj/20.18.5129

650. Schleicher S, Boekhoff I, Arriza J, Lefkowitz RJ, Breer H (1993) A beta-adrenergic receptor kinase-like enzyme is involved in olfactory signal termination. *Proc Natl Acad Sci USA* 90: 1420-1424. DOI: 10.1073/pnas.90.4.1420

651. Watanabe H, Xu J, Bengra C, Jose PA, Felder RA (2002) Desensitization of human renal D1 dopamine receptors by G protein-coupled receptor kinase 4. *Kidney Int* 62: 790-798. DOI: 10.1046/j.1523-1755.2002.00525.x

652. Overland AC, Kitto KF, Chabot-Dore AJ, Rothwell PE, Fairbanks CA, et al. (2009) Protein kinase c mediates the synergistic interaction between agonists acting at alpha(2)-adrenergic and delta-opioid receptors in spinal cord. *J Neurosci* 29: 13264-13273. DOI: 10.1523/JNEUROSCI.1907-09.2009

653. Rios C, Gomes I, Devi LA (2004) Interactions between delta opioid receptors and alpha-adrenoceptors. *Clin Exp Pharmacol Physiol* 31: 833-836. DOI: 10.1111/j.1440-1681.2004.04076.x

654. Ramsay D, Kellett E, McVey M, Rees S, Milligan G (2002) Homo- and hetero-oligomeric interactions between G-protein-coupled receptors in living cells monitored by two variants of bioluminescence resonance energy transfer (BRET): hetero-oligomers between receptor subtypes form more efficiently than between less closely related sequences. *Biochem J* 365: 429-440. DOI: 10.1042/BJ20020251

655. Jordan BA, Trapaidze N, Gomes I, Nivarthi R, Devi LA (2001) Oligomerization of opioid receptors with beta 2-adrenergic receptors: a role in trafficking and mitogen-activated protein kinase activation. *Proc Natl Acad Sci USA* 98: 343-348.

656. Rozenfeld R, Bushlin I, Gomes I, Tzavaras N, Gupta A, et al. (2012) Receptor heteromerization expands the repertoire of cannabinoid signaling in rodent neurons. *PLoS One* 7: e29239. DOI: 10.1371/journal.pone.0029239

657. Rios C, Gomes I, Devi LA (2006) mu opioid and CB1 cannabinoid receptor interactions: reciprocal inhibition of receptor signaling and neuritogenesis. *Br J Pharmacol* 148: 387-395. DOI: 10.1038/sj.bjp.0706757

658. Pello OM, Martinez-Munoz L, Parrillas V, Serrano A, Rodriguez-Frade JM, et al. (2008) Ligand stabilization of CXCR4/delta-opioid receptor heterodimers reveals a mechanism for immune response regulation. *Eur J Immunol* 38: 537-549. DOI: 10.1002/eji.200737630

659. Ambrose-Lanci LM, Peiris NB, Unterwald EM, Van Bockstaele EJ (2008) Cocaine withdrawal-induced trafficking of delta-opioid receptors in rat nucleus accumbens. *Brain Res* 1210: 92-102. DOI: 10.1016/j.brainres.2008.02.105

660. Ambrose LM, Gallagher SM, Unterwald EM, Van Bockstaele EJ (2006) Dopamine-D1 and delta-opioid receptors co-exist in rat striatal neurons. *Neurosci Lett* 399: 191-196. DOI: 10.1016/j.neulet.2006.02.027

661. Jordan BA, Devi LA (1999) G-protein-coupled receptor heterodimerization modulates receptor function. *Nature* 399: 697-700. DOI: 10.1038/21441

662. Kabli N, Martin N, Fan T, Nguyen T, Hasbi A, et al. (2010) Agonists at the delta-opioid receptor modify the binding of micro-receptor agonists to the micro-delta receptor hetero-oligomer. *Br J Pharmacol* 161: 1122-1136. DOI: 10.1111/j.1476-5381.2010.00944.x

663. Fan T, Varghese G, Nguyen T, Tse R, O'Dowd BF, et al. (2005) A role for the distal carboxyl tails in generating the novel pharmacology and G protein activation profile of mu and delta opioid receptor hetero-oligomers. *J Biol Chem* 280: 38478-38488. DOI: 10.1074/jbc.M505644200

664. Charles AC, Mostovskaya N, Asas K, Evans CJ, Dankovich ML, et al. (2003) Coexpression of delta-opioid receptors with micro receptors in GH3 cells changes the functional response to micro agonists from inhibitory to excitatory. *Mol Pharmacol* 63: 89-95. DOI: 10.1124/mol.63.1.89

665. Law PY, Erickson-Herbrandson LJ, Zha QQ, Solberg J, Chu J, et al. (2005) Heterodimerization of mu- and delta-opioid receptors occurs at the cell surface only and requires receptor-G protein interactions. *J Biol Chem* 280: 11152-11164. DOI: 10.1074/jbc.M500171200

666. Milan-Lobo L, Whistler JL (2011) Heteromerization of the mu- and delta-opioid receptors produces ligand-biased antagonism and alters mu-receptor trafficking. *J Pharmacol Exp Ther* 337: 868-875. DOI: 10.1124/jpet.111.179093

667. Li Y, Chen J, Bai B, Du H, Liu Y, et al. (2012) Heterodimerization of human apelin and kappa opioid receptors: roles in signal transduction. *Cell Signal* 24: 991-1001. DOI: 10.1016/j.cellsig.2011.12.012

668. Jordan BA, Gomes I, Rios C, Filipovska J, Devi LA (2003) Functional interactions between mu opioid and alpha 2A-adrenergic receptors. *Mol Pharmacol* 64: 1317-1324. DOI: 10.1124/mol.64.6.1317

669. Vilardaga JP, Nikolaev VO, Lorenz K, Ferrandon S, Zhuang Z, et al. (2008) Conformational cross-talk between alpha2A-adrenergic and mu-opioid receptors controls cell signaling. *Nat Chem Biol* 4: 126-131. DOI: 10.1038/nchembio.64

670. Zhang YQ, Limbird LE (2004) Hetero-oligomers of alpha2A-adrenergic and mu-opioid receptors do not lead to transactivation of G-proteins or altered endocytosis profiles. *Biochem Soc Trans* 32: 856-860. DOI: 10.1042/BST0320856

671. Hojo M, Sudo Y, Ando Y, Minami K, Takada M, et al. (2008) mu-Opioid receptor forms a functional heterodimer with cannabinoid CB1 receptor: electrophysiological and FRET assay analysis. *J Pharmacol Sci* 108: 308-319. DOI: 10.1254/jphs.08244FP

672. Chen C, Li J, Bot G, Szabo I, Rogers TJ, et al. (2004) Heterodimerization and cross-desensitization between the mu-opioid receptor and the chemokine CCR5 receptor. *Eur J Pharmacol* 483: 175-186. DOI: 10.1016/j.ejphar.2003.10.033

673. Suzuki S, Chuang LF, Yau P, Doi RH, Chuang RY (2002) Interactions of opioid and chemokine receptors: Oligomerization of mu, kappa, and delta with CCR5 on immune cells. *Exp Cell Res* 280: 192-200. DOI: 10.1006/excr.2002.5638

674. Liu XY, Liu ZC, Sun YG, Ross M, Kim S, et al. (2011) Unidirectional cross-activation of GRPR by MOR1D uncouples itch and analgesia induced by opioids. *Cell* 147: 447-458. DOI: 10.1016/j.cell.2011.08.043

675. Chakrabarti S, Liu NJ, Gintzler AR (2010) Formation of mu-/kappa-opioid receptor heterodimer is sex-dependent and mediates female-specific opioid analgesia. *Proc Natl Acad Sci USA* 107: 20115-20119. DOI: 10.1073/pnas.1009923107

676. Pan YX, Bolan E, Pasternak GW (2002) Dimerization of morphine and orphanin FQ/nociceptin receptors: generation of a novel opioid receptor subtype. *Biochem Biophys Res Commun* 297: 659-663. DOI: 10.1016/S0006-291X(02)02258-1

677. Pfeiffer M, Koch T, Schroder H, Laugsch M, Hollt V, et al. (2002) Heterodimerization of somatostatin and opioid receptors cross-modulates phosphorylation, internalization, and desensitization. *J Biol Chem* 277: 19762-19772. DOI: 10.1074/jbc.M110373200

678. Carriba P, Ortiz O, Patkar K, Justinova Z, Stroik J, et al. (2007) Striatal adenosine A(2A) and Cannabinoid CB1 receptors form functional heteromeric complexes that mediate the motor effects of Cannabinoids. *Neuropsychopharmacol* 32: 2249-2259. DOI: 10.1038/sj.npp.1301375

679. Hudson BD, Hebert TE, Kelly MEM (2010) Physical and functional interaction between CB1 cannabinoid receptors and beta(2)-adrenoceptors. *Br J Pharmacol* 160: 627-642. DOI: 10.1111/j.1476-5381.2010.00681.x

680. Callen L, Moreno E, Barroso-Chinea P, Moreno-Delgado D, Cortes A, et al. (2012) Cannabinoid receptors CB1 and CB2 form functional heteromers in brain. *J Biol Chem* 287: 20851-20865. DOI: 10.1074/jbc.M111.335273

681. Giuffrida A, Parsons LH, Kerr TM, Rodriguez de Fonseca F, Navarro M, et al. (1999) Dopamine activation of endogenous cannabinoid signaling in dorsal striatum. *Nat Neurosci* 2: 358-363. DOI: 10.1038/7268

682. Glass M, Felder CC (1997) Concurrent stimulation of cannabinoid CB1 and dopamine D2 receptors augments cAMP accumulation in striatal neurons: evidence for a Gs linkage to the CB1 receptor. *J Neurosci* 17: 5327-5333.

683. Jarrahian A, Watts VJ, Barker EL (2004) D2 dopamine receptors modulate Galpha-subunit coupling of the CB1 cannabinoid receptor. *J Pharmacol Exp Ther* 308: 880-886. DOI: 10.1124/jpet.103.057620

684. Kearn CS, Blake-Palmer K, Daniel E, Mackie K, Glass M (2005) Concurrent stimulation of cannabinoid CB1 and dopamine D2 receptors enhances heterodimer formation: a mechanism for receptor cross-talk? *Mol Pharmacol* 67: 1697-1704. DOI: 10.1124/mol.104.006882

685. Ward RJ, Pediani JD, Milligan G (2011) Heteromultimerization of Cannabinoid CB1 Receptor and Orexin OX1 Receptor Generates a Unique Complex in Which Both Protomers Are Regulated by Orexin A. *J Biol Chem* 286: 37414-37428. DOI: 10.1074/jbc.M111.287649

686. Hilairet S, Bouaboula M, Carriere D, Le Fur G, Casellas P (2003) Hypersensitization of the orexin 1 receptor by the CB1 receptor - Evidence for cross-talk blocked by the specific CB1 antagonist, SR141716. *J Biol Chem* 278: 23731-23737. DOI: 10.1074/jbc.M212369200

687. Urizar E, Yano H, Kolster R, Gales C, Lambert N, et al. (2011) CODA-RET reveals functional selectivity as a result of GPCR heteromerization. *Nat Chem Biol* 7: 624-630. DOI: 10.1038/nchembio.623

688. Lee SP, So CH, Rashid AJ, Varghese G, Cheng R, et al. (2004) Dopamine D1 and D2 receptor Co-activation generates a novel phospholipase C-mediated calcium signal. *J Biol Chem* 279: 35671-35678. DOI: 10.1074/jbc.M401923200

689. Rashid AJ, So CH, Kong MM, Furtak T, El-Ghundi M, et al. (2007) D1-D2 dopamine receptor heterooligomers with unique pharmacology are coupled to rapid activation of Gq/11 in the striatum. *Proc Natl Acad Sci USA* 104: 654-659. DOI: 10.1073/pnas.0604049104

690. Hasbi A, Fan T, Alijaniaram M, Nguyen T, Perreault ML, et al. (2009) Calcium signaling cascade links dopamine D1-D2 receptor heteromer to striatal BDNF production and neuronal growth. *Proc Natl Acad Sci USA* 106: 21377-21382. DOI: 10.1073/pnas.0903676106

691. Verma V, Hasbi A, O'Dowd BF, George SR (2010) Dopamine D1-D2 receptor heteromer-mediated calcium release is desensitized by D1 receptor occupancy with or without signal activation dual functional regulation by g protein-coupled receptor kinase 2. *J Biol Chem* 285: 35092-35103. DOI: 10.1074/jbc.M109.088625

692. Perreault ML, Hasbi A, Alijaniaram M, Fan T, Varghese G, et al. (2010) The dopamine D1-D2 receptor heteromer localizes in dynorphin/enkephalin neurons increased high affinity state following amphetamine and in schizophrenia. *J Biol Chem* 285: 36625-36634. DOI: 10.1074/jbc.M110.159954

693. Fiorentini C, Busi C, Gorruso E, Gotti C, Spano P, et al. (2008) Reciprocal regulation of dopamine D1 and D3 receptor function and trafficking by heterodimerization. *Mol Pharmacol* 74: 59-69. DOI: 10.1124/mol.107.043885

694. Gines S, Hillion J, Torvinen M, Le Crom S, Casado V, et al. (2000) Dopamine D1 and adenosine A1 receptors form functionally interacting heteromeric complexes. *Proc Natl Acad Sci USA* 97: 8606-8611. DOI: 10.1073/pnas.150241097

695. Maggio R, Millan MJ (2010) Dopamine D2-D3 receptor heteromers: pharmacological properties and therapeutic significance. *Curr Opin Pharmacol* 10: 100-107. DOI: 10.1016/j.coph.2009.10.001

696. Scarselli M, Novi F, Schallmach E, Lin R, Baragli A, et al. (2001) D2/D3 dopamine receptor heterodimers exhibit unique functional properties. *J Biol Chem* 276: 30308-30314. DOI: 10.1074/jbc.M102297200

697. Hillion J, Canals M, Torvinen M, Casado V, Scott R, et al. (2002) Coaggregation, cointernalization, and codesensitization of adenosine A2A receptors and dopamine D2 receptors. *J Biol Chem* 277: 18091-18097. DOI: 10.1074/jbc.M107731200

698. Trincavelli ML, Daniele S, Orlandini E, Navarro G, Casado V, et al. (2012) A new D-2 dopamine receptor agonist allosterically modulates A(2A) adenosine receptor signalling by interacting with the A(2A)/D-2 receptor heteromer. *Cell Signal* 24: 951-960. DOI: 0.1016/j.cellsig.2011.12.018

699. Canals M, Marcellino D, Fanelli F, Ciruela F, de Benedetti P, et al. (2003) Adenosine A(2A)-dopamine D2 receptor-receptor heteromerization - Qualitative and quantitative assessment by fluorescence and bioluminescence energy transfer. *J Biol Chem* 278: 46741-46749. DOI: 10.1074/jbc.M306451200

700. Kamiya T, Saitoh O, Yoshioka K, Nakata H (2003) Oligomerization of adenosine A(2A) and dopamine D-2 receptors in living cells. *Biochem Biophys Res Commun* 306: 544-549. DOI: 10.1016/S0006-291X(03)00991-4

701. Fuxe K, Canals M, Torvinen M, Marcellino D, Terasmaa A, et al. (2007) Intramembrane receptor-receptor interactions: a novel principle in molecular medicine. *J Neural Transm* 114: 49-75. DOI: 10.1007/s00702-006-0589-0

702. Kern A, Albarran-Zeckler R, Walsh HE, Smith RG (2012) Apo-ghrelin receptor forms heteromers with DRD2 in hypothalamic neurons and is essential for anorexigenic effects of DRD2 agonism. *Neuron* 73: 317-332. DOI: 10.1016/j.neuron.2011.10.038

703. Koschatzky S, Tschammer N, Gmeiner P (2011) Cross-Receptor Interactions between Dopamine D(2L) and Neurotensin NTS(1) Receptors Modulate Binding Affinities of Dopaminergics. *ACS Chem Neurosci* 2: 308-316. DOI: 10.1021/cn200020y

704. Thibault D, Albert PR, Pineyro G, Trudeau LE (2011) Neurotensin triggers dopamine D2 receptor desensitization through a protein kinase C and beta-arrestin1-dependent mechanism. *J Biol Chem* 286: 9174-9184. DOI: 10.1074/jbc.M110.166454

705. Romero-Fernandez W, Borroto-Escuela DO, Agnati LF, Fuxe K (2012) Evidence for the existence of dopamine d2-oxytocin receptor heteromers in the ventral and dorsal striatum with facilitatory receptor-receptor interactions. *Mol Psychiatry* doi: 10.1038/mp.2012.103 (Epub ahead of print) DOI: 10.1038/mp.2012.103

706. Baragli A, Alturaihi H, Watt HL, Abdallah A, Kumar U (2007) Heterooligomerization of human dopamine receptor 2 and somatostatin receptor 2 Co-immunoprecipitation and fluorescence resonance energy transfer analysis. *Cell Signal* 19: 2304-2316. DOI: 10.1016/j.cellsig.2007.07.007

707. Rocheville M, Lange DC, Kumar U, Patel SC, Patel RC, et al. (2000) Receptors for dopamine and somatostatin: formation of hetero-oligomers with enhanced functional activity. *Science* 288: 154-157. DOI: 10.1126/science.288.5463.154

708. Albizu L, Holloway T, Gonzalez-Maeso J, Sealfon SC (2011) Functional crosstalk and heteromerization of serotonin 5-HT2A and dopamine D2 receptors. *Neuropharmacology* 61: 770-777. DOI: 10.1016/j.neuropharm.2011.05.023

709. Gonzalez S, Moreno-Delgado D, Moreno E, Perez-Capote K, Franco R, et al. (2012) Circadian-related heteromerization of adrenergic and dopamine D(4) receptors modulates melatonin synthesis and release in the pineal gland. *PLoS Biol* 10: e1001347. DOI: 10.1371/journal.pbio.1001347

710. Benjannet S, Rondeau N, Day R, Chretien M, Seidah NG (1991) PC1 and PC2 are proprotein convertases capable of cleaving proopiomelanocortin at distinct pairs of basic residues. *Proc Natl Acad Sci USA* 88: 3564-3568. DOI: 10.1073/pnas.88.9.3564

711. Thomas G (2002) Furin at the cutting edge: from protein traffic to embryogenesis and disease. *Nat Rev Mol Cell Biol* 3: 753-766. DOI: 10.1038/nrm934

712. Seidah NG, Mayer G, Zaid A, Rousselet E, Nassoury N, et al. (2008) The activation and physiological functions of the proprotein convertases. *Int J Biochem Cell Biol* 40: 1111-1125. DOI: 10.1016/j.biocel.2008.01.030

713. Steiner DF (1998) The proprotein convertases. *Curr Opin Chem Biol* 2: 31-39. DOI: 10.1016/S1367-5931(98)80033-1

714. Turner AJ, Matsas R, Kenny AJ (1985) Are there neuropeptide-specific peptidases? *Biochem Pharmacol* 34: 1347-1356. DOI: 10.1016/0006-2952(85)90669-0

715. Medeiros MS, Turner AJ (1994) Post-secretory processing of regulatory peptides: the pancreatic polypeptide family as a model example. *Biochimie* 76: 283-287. DOI: 10.1016/0300-9084(94)90159-7

716. Cummins PM, O'Connor B (1998) Pyroglutamyl peptidase: an overview of the three known enzymatic forms. *Biochim Biophys Acta* 1429: 1-17. DOI: 10.1016/S0167-4838(98)00248-9

717. Mentlein R (1999) Dipeptidyl-peptidase IV (CD26)--role in the inactivation of regulatory peptides. *Regul Pept* 85: 9-24. DOI: 10.1016/S0167-0115(99)00089-0

718. Raffin-Sanson ML, de Keyzer Y, Bertagna X (2003) Proopiomelanocortin, a polypeptide precursor with multiple functions: from physiology to pathological conditions. *Eur J Endocrinol* 149: 79-90. DOI: 10.1530/eje.0.1490079

719. Kieffer BL (1999) Opioids: first lessons from knockout mice. *Trends Pharmacol Sci* 20: 19-26. DOI: 10.1016/S0165-6147(98)01279-6

720. Mariel B (2000) Vasopressin Receptors. *Trends Endocrinol Metab* 11: 406-410. DOI: 10.1016/S1043-2760(00)00304-0

721. Boyd ST (2006) The endocannabinoid system. Pharmacotherapy 26: 218S-221S. DOI: 10.1592/phco.26.12part2.218S

722. Maldonado R, Valverde O, Berrendero F (2006) Involvement of the endocannabinoid system in drug addiction. *Trends Neurosci* 29: 225-232. DOI: 10.1016/j.tins.2006.01.008

723. Porter AC, Felder CC (2001) The endocannabinoid nervous system: Unique opportunities for therapeutic intervention. *Pharmacol Ther* 90: 45-60. DOI: 10.1016/S0163-7258(01)00130-9

724. Miller LK, Devi LA (2011) The highs and lows of cannabinoid receptor expression in disease: mechanisms and their therapeutic implications. *Pharmacol Rev* 63: 461-470. DOI: 10.1124/pr.110.003491